Teaching Mathematics to Children with Special Needs

Fredricka K. Reisman
University of Georgia

Samuel H. Kauffman, M.D.

CHARLES E. MERRILL PUBLISHING COMPANY
A Bell & Howell Company
Columbus   Toronto   London   Sydney

Published by
CHARLES E. MERRILL PUBLISHING CO.
A Bell & Howell Company
Columbus, Ohio 43216

This book was set in Souvenir.
Production Editor: Judith Rose Sacks
Cover Design Coordination: Will Chenoweth
Text Designer: Amato Prudente

to Lisa with our love

Library of Congress Catalog Card Number: 79-92854
International Standard Book Number: 0–675-0-8175–0
Printed in the United States of America
2 3 4 5 6 7 8 9 10 — 85 84 83 82

Acknowledgments

The authors are grateful to the following colleagues for providing critical review and suggestions on earlier drafts of the manuscript: Henry Kepner, Phillip Peak, Jon Engelhardt, Floyd Hudson, and Andrew Shotick. Students using a prepublication manuscript who were particularly helpful included Brenda Floyd, Jack Riley, and Jan Stewart. Individuals at Charles E. Merrill were extremely helpful. The authors are indebted to their production editor, Judy Sacks, whose excellence and scholarship in editing the manuscript was surpassed only by her graciousness and delightful creative nature. Marianne Taflinger and Vicki Knight both made our endeavors so pleasant in their unique ways. To Tom Hutchinson, Executive Editor, a very special thank you for all of your expertise, support, and creative suggestions.

Thanks are also due those who typed the manuscript, especially Cindy Green, Wanda Callaham, Elizabeth Platt, Joyce Davis and Cami Damien.

How to Use the Diagnostic Guide By Mathematics Topic (Appendix H)

Step 1: Select the mathematics topic or content to be taught.

Step 2: Find the mathematics content to be taught under the column entitled *Mathematics Topic.*

Step 3: Skim the generic factors that influence learning this topic listed under the column entitled *Related Special Educational Needs.*

Step 4: Decide whether one or more of the special educational needs related to the specific topic under consideration are characteristic of the student or group of students to whom you wish to teach the topic.

Step 5: Use the most relevant teaching activities listed under the column entitled *Suggested Instructional Strategy (IS).*

Following is an excerpt from the *Diagnostic Guide* to serve as an illustration of applying Steps 1–5:

Mathematics Topic	Related Special Educational Needs	Suggested Instructional Strategy (IS)
5.3.0 Topological relationships	1.8.0 Difficulty forming relationships	5.3.1 Reinforce attention to a relevant dimension
	1.9.0 Difficulty attending to salient aspects of situations	5.3.2 Point out relevant relationships
	2.0.0 Psychomotor difficulties	5.3.3 Incorporate movement to develop order
	3.1.0 Physical impairment	11.1.5 Use multisensory approaches
	3.3.0 Sensory limitation	11.2.2 Use nonvisual cues

Contents

Foreword   **xi**

Preface   **xii**

Introduction: How to Use This Book   **xiii**

Section I   Generic Factors that Influence Learning Mathematics   **1**

      Chapter 1   Cognitive Factors that Influence Learning Mathematics   **3**

      1.1.0   Rate and amount of learning compared to age peers   **3**

      1.2.0   Speed of learning related to specific content   **4**

      1.3.0   Ability to retain information   **5**

      1.4.0   Need for repetition   **5**

      1.5.0   Verbal skills   **6**

      1.6.0   Ability to learn arbitrary associations and symbol systems   **7**

      1.7.0   Size of vocabulary compared with peers   **10**

      1.8.0   Ability to form relationships, concepts, and generalizations   **10**

      1.9.0   Ability to attend to salient aspects of a situation   **11**

      1.10.0 Use of problem-solving strategies   **12**

      1.11.0 Ability to make decisions and judgments   **13**

      1.12.0 Ability to draw inferences and conclusions and to hypothesize   **13**

      1.13.0 Ability to abstract and cope with complexity   **13**

Chapter 2 Psychomotor Factors that Influence Learning Mathematics   20

2.1.0 Disorders of visual perception   **22**
2.2.0 Disorders of auditory perception   **24**
2.3.0 Rules of general language and mathematics   **26**

Chapter 3 Physical and Sensory Factors that Influence Learning Mathematics   30

3.1.0 Physical impairments   **31**
3.2.0 Low-vitality and fatigue conditions   **32**
3.3.0 Sensory limitation   **32**

Chapter 4 Social and Emotional Factors that Influence Learning Mathematics   36

4.1.0 Hyperactivity, distractibility, impulsivity   **40**
4.2.0 Aggressiveness   **41**
4.3.0 Withdrawal, immaturity, inadequacy   **41**
4.4.0 Deficiencies in moral development   **41**

Section II Instructional Strategies   45

Chapter 5 Instructional Strategies Based on Cognitive Needs: Basic Relationships and Arbitrary Associations   51

5.0.0 Prenumber relationships   **51**
5.1.0 Sequencing   **52**
5.2.0 One-to-one correspondence   **60**
5.3.0 Topological relationships   **62**
5.4.0 Arbitrary associations   **64**

Chapter 6 Instructional Strategies Based on Cognitive Needs: Lower-Level Generalizations   73

6.1.0 Duality   **73**
6.2.0 Equivalence of sets   **75**
6.3.0 Greater than-less than matching   **77**
6.4.0 Sorting   **80**

6.5.0 Many-to-one and one-to-many matching   **83**
6.6.0 Seriation   **84**
6.7.0 Judging and estimating   **89**

Chapter 7   Instructional Strategies Based on Cognitive Needs: Concepts   **93**

Chapter 8   Instructional Strategies Based on Cognitive Needs: Higher-Level Relationships   **101**

8.1.0 Transformations   **101**
8.2.0 Set operations and relationships   **109**
8.3.0 Number operations and relationships   **117**
8.4.0 Mathematics as a language of relationships   **123**

Chapter 9   Instructional Strategies Based on Cognitive Needs: Higher-Level Generalizations   **138**

9.1.0 Basic facts   **138**
9.2.0 Axioms   **148**
9.3.0 Place value   **152**
9.4.0 Computation   **159**
9.5.0 Seriation extended   **165**
9.6.0 Solving word problems   **167**
9.7.0 Cause-effect extended   **170**
9.8.0 Topological equivalence   **170**
9.9.0 Projective geometry   **171**
9.10.0 Nonmetric geometry   **171**
9.11.0 Metric geometry   **171**
9.12.0 Network theory   **171**
9.13.0 Probability, statistics, graphs   **171**
9.14.0 Applied mathematics-measurement   **171**
9.15.0 Applied mathematics-economics   **171**

Chapter 10   Instructional Strategies Based on Psychomotor Needs   **175**

10.1.0 Psychomotor disabilities   **175**

Chapter 11 Instructional Strategies Based on Sensory and Physical Needs   184

11.1.0 Impaired hearing **184**
11.2.0 Impaired vision **188**
11.3.0 Physical impairment **189**
11.4.0 Low-vitality and fatigue conditions **192**

Chapter 12 Instructional Strategies Based on Social and Emotional Needs   196

12.1.0 Hyperactivity, distractibility, and impulsivity **196**
12.2.0 Aggressiveness **198**
12.3.0 Withdrawal, immaturity and inadequacy **198**
12.4.0 Deficiencies in moral development **199**

Section III Implementation   203

Chapter 13 Interviewing Techniques   205

Functions of interviews **205**
Types of questions **206**
Techniques for dealing with communication blocks **207**

Chapter 14 Mainstreaming and the Individualized Mathematics Program (IEPM)   212

Mainstreaming **212**
Developing an IEPM **214**
Functions of the IEPM committee **215**
Annual goals **215**
Short-term goals and behavioral objectives **219**
Evaluation of goals and objectives **221**
IEP formats **221**

Chapter 15 "How Can I Help My Child?" Changing Roles of Parents and Teachers   226

Parents' rights under P.L. 94-142 **227**

Appendix A: Teacher's Review of Selected Mathematics
Content   231

Appendix B: Reisman-Kauffman Interview Checklist   237

Appendix C: Instructional Strategies (IS) for Specific
Learning Disabilities in Mathematics
(SLDM) Checklist   240

Appendix D: Developmental Mathematics Curriculum —
Screening Checklist   243

Appendix E: Some Trouble Spots in the Primary
Grades Mathematics Curriculum   254

Appendix F: Case Studies and IEPMs   258

Appendix G: Cross-Reference Outline by Instructional
Strategies   270

Appendix H: Diagnostic Guide by Mathematics Topic   296

Author Index   311

Subject Index   315

Foreword

Today, the law guarantees each youngster an equal educational opportunity and a free and appropriate education through PL 94–142, *The Education of All Handicapped Children Act of 1975,* PL 95–561 (Title IX–A), *The Gifted and Talented Children's Education Act of 1978,* and other statutes and judicial rulings. Our task is to translate these guarantees into reality — to be sure that each pupil receives the rights and privileges which he or she deserves as a citizen. To carry out this task, we need knowledge and understanding about our pupils — how they learn and how we can teach them best. Drs. Reisman and Kauffman have contributed some of that knowledge and understanding in this book. In it, they present procedures for teaching mathematics to exceptional pupils, both the handicapped and the talented. They emphasize generic factors important in learning and thereby encourage the consideration of core influences in learning. Further, by arranging chapters hierarchically according to the types of cognitive processing operating in the respective types of mathematics performance, they stress task analyses of mathematics topics and identification of prerequisite relationships, concepts, or generalizations. Also, the authors point out trouble spots in the early-grades mathematics curriculum which appear during clinical teaching of mathematics, and they provide practical checklists useful with pupils who have special needs.

These ideas supply a previously unmet need in our work with exceptional pupils. They can be valuable tools in helping us facilitate pupils' growth and attainment of rights as citizens. I recommend this book highly to those responsible for educating exceptional pupils — their families and their teachers.

Kathryn A. Blake

Head, Division for the Education
of Exceptional Children

University of Georgia

Preface

In 1961, the Educational Policies Commission of the National Education Association recommended that the "ten rational processes" be made the central objective of education in the United States. The rational processes were defined as: recalling and imagining, classifying and generalizing, comparing and evaluating, analyzing and synthesizing, and deducing and inferring. Since 1961, many textbook writers and teachers have continued to ignore this recommendation. Many, however, have given serious attention to the development of these "ten rational processes," and studies have shown that today's students are more skilled in abstract reasoning and creative thinking than their counterparts in 1960.

Since 1961, psychologists have identified and studied a number of additional rational processes. They have also given attention to important, practical thinking skills that go beyond the rational processes. These processes have been labeled as: intuitive thinking, suprarational thinking, creative thinking, and the like. In real life, these processes have long been recognized as important, despite the fact that they have rarely been valued in formal schooling. In this book, Reisman and Kauffman have given attention not only to the "ten rational processes" but also to other important rational processes and those practical processes that go beyond rational, logical thinking. They show how mathematics teachers can encourage the development of these processes, and how this can be done during the preschool and primary school years. Furthermore, the kind of learning and teaching they describe sounds exciting and full of fun. I have the feeling that such teaching would reduce the fear and failure frequently associated with the learning of mathematics.

A big advantage of the kinds of learning and teaching described by Reisman and Kauffman is that they are designed to facilitate learning among all children, including the handicapped and the gifted. The authors do this by showing how learning strengths can be identified and used in developing effective, powerful teaching strategies. Furthermore, all children are taught through the uses of all modalities — visual, auditory, kinesthetic, and tactual. In a sense, the theme of the book is reflected in the idea that children "learn to move and move to learn." They show how children can learn transformational thinking through mathematics, and they learn mathematics through transformational thinking. Children learn creative thinking skills through mathematics, and they learn mathematics through doing creative thinking.

This is an exciting and provocative book and will be useful to almost any teacher who is interested in making teaching an interesting and exciting adventure.

E. Paul Torrance

University of Georgia

January, 1980

Introduction
How to Use This Book

Section I

Mathematics instruction for students with special educational needs is a part of the changing role for regular classroom teachers involved in mainstreaming and for resource teachers who previously have been trained to emphasize reading and language skills. In order to teach mathematics to children with special educational problems, it helps to know what generic factors influence learning in general. Generic factors are those conditions that may either enhance or inhibit learning. They are general influences that affect how humans learn, and therefore they must be considered in order to differentiate instruction to accommodate learning needs.

Differential instruction involves selecting curriculum (for example, mathematics content) that is relevant to a child's educational development. Differential instruction also means developing methods and materials to accommodate the learner's unique needs. Persons develop cognitively at varying rates. Some are able to grasp new ideas very quickly, and some take a long time for understanding to come about. Some can comprehend complex relationships; some can understand only the simplest. Some persons can remember ideas easily and need a minimum of repetition; some need a great deal of repetition and practice. Some can learn *incidentally*, picking out the important aspect of a situation on their own, while others need a lot of structure to become aware of what it is they must notice.

Some students use cues to help them notice and remember salient attributes of a learning situation, while others need to be told to look for designated cues. Some students may have abundant energy and stick with tasks for long periods of time. Some may fatigue quickly due to a physical condition such as diabetes or epilepsy or because thinking is hard work and uses a lot of their energy. For these students, the teacher must adjust the amount and pace of instruction. Some thrive in a colorful, busy environment; some become overly excited and distracted easily and need a calm, sterile environment free of distracting stimuli.

For some students, a combination of impairments of generic factors may be influencing learning. A slowly developing cognitive rate, poor memory, inability to engage in abstract reasoning, inability to cope with complexity, and the confounding effect of a visual and/or auditory impairment would have a greater impact on learning than if no sensory handicap is involved. On the other hand, for a student who is developing at a very rapid cognitive rate, who is well adjusted socially and emotionally, but who is confined to a wheelchair because of a birth defect, the impact of the physical impairment on learning mathematics may be minimal. Section I of this book thus serves to introduce the basic factors that influence learning in general.

Section II

Section II presents instructional strategies (ISs) and discussion of curriculum selection to accommodate extreme conditions of the generic factors that influence learning. "Extreme" conditions mean both handicaps and talents. A major goal of this book is to serve as an instructional tool for persons involved in the mathematics education of the handicapped and the talented, including those students in the mainstream. Many of the instructional strategies presented in Section II may be used by both special and regular educators, either in resource rooms or for a child who is mainstreamed in a regular classroom setting for all or most of the day. Some of the ISs require special training and competence — for example, the use of Braille, sign language, or in-depth knowledge of behavior modification theory and techniques.

The chapters that comprise Section II relate to chapters in Section I. For example, Chapters 5 through 9 include instructional strategies that accommodate cognitive factors discussed in Chapter 1. Suggestions for methods, materials, and appropriate topics in light of psychomotor, physical and sensory needs are included in Chapters 10 and 11. These modifications relate to generic factors discussed in Chapters 2 and 3, respectively. Instructional strategies in Chapter 12 are particularly appropriate for social and emotional generic factors discussed in Chapter 4.

Section III

Section III focuses on implementation. A goal of the authors is to help both special and regular educators begin to develop competencies for creating Individual Education Programs in Mathematics (IEPMs). Chapter 13 describes interview techniques; Chapter 14 deals with steps in developing an IEP in mathematics, (an IEPM); and in Chapter 15, issues of parents' rights and parent - teacher cooperative planning are discussed.

Aids to the Reader

In addition to serving as an instructional tool for college coursework, the book contains practical information for developing mathematics instruction for special educational needs students. Therefore, the Appendices include checklists that may be used directly with students, trouble spots in mathematics that many students encounter, and a very brief basic review of mathematics content particularly for those whose programs of study may not have included much mathematics. Appendix F contains additional case study summaries that are presented as models for helping the reader to tie together results of various assessments and for developing IEPMs.

Appendix G, the *Cross-Reference Outline by Instructional Strategies,* and Appendix H, *Diagnostic Guide by Mathematics Topic,* contain the same information but are organized in different ways and serve different purposes. The *Diagnostic Guide* (shaded in gray for easy reference) focuses on mathematics topics, while the *Cross-Reference Outline* is organized by instructional strategies (ISs). If you know that you want to teach a particular mathematics topic, find the topic in the first column of the *Diagnostic Guide*; read which of the inhibiting generic factors most affect its acquisition; and then select one or more of the ISs that are suggested. On the other hand, if you want to use a particular IS with topics other than the one suggested in the chapter, turn to the *Cross-Reference Outline* and identify other topics for which it may be appropriate.

The authors wish to emphasize that competence in adapting mathematics instruction to the special educational needs of a student does not develop overnight. The task is challenging and may seem overwhelming, especially for those with a minimum of teaching experience. Interdisciplinary cooperation is necessary as the boundaries between regular and special education are lessened.

Generic Factors That Influence Learning Mathematics

SECTION I

Educators are no longer focusing upon methods of grouping children for teaching purposes based mainly on identifiable physical or intellectual weaknesses. The shift is away from the medical model of diagnosis and remediation of weaknesses to the more operational function model—an educative model that looks to generic factors that influence learning. For example, knowing about visual acuity or etiological factors of a child's blindness offers less hope for educational success than does describing how well the child makes use of what vision she or he has. In fact, it has been shown (Barraga, 1964) that giving visually impaired children strategies for seeing by capitalizing on their strengths has improved their visual efficiency. When the teacher is aware of such generic factors and uses an exceptional child's strengths to circumvent weaknesses, learning can occur.

The book is structured upon four main assumptions:

1. There are identifiable generic factors—cognitive, physical, social, and emotional—that influence learning mathematics.
2. Mathematics has a developmental nature that serves as a structure for curriculum and pedogogical decisions.
3. Observation of the impact of a handicap or talent on learning provides information for developing mathematics instruction.
4. Teachers can to some degree circumvent deficits such as slow rate of mental development, difficulty in retaining information, sensory

limitation, or inability to attend, through the use of instructional strategies.

The first assumption underlies Section I. Generic factors presented in Chapters 1, 2, 3, and 4 form the basis for strategies in mathematics instruction in Section II.

Reference

Barraga, N. Increased visual behavior in low vision children. *Research series no. 13.* New York: American Foundation for the Blind, 1964.

Cognitive Factors That Influence Learning Mathematics

1

For a teacher, it does not necessarily help to know if a child is profoundly, severely, or moderately retarded, or "gifted." These labels do not tell what mathematics the child could learn, and they do not tell how to teach. Instead, it is helpful to know that one of the core influences on learning is the rate at which one acquires knowledge. Knowing the scope of mathematics that a child may be expected to learn in a given span of time provides guidance for selecting those topics that are most relevant for a particular child. Each of the generic or core factors discussed in this chapter may be applied to intellectually handicapped as well as talented learners. These generic factors in their extreme have a profound impact on a child's learning mathematics. The child's ability to see relationships, to engage in abstract thinking, to use symbol systems and to problem-solve will affect his or her acquisition of mathematics ideas.

Cognitive considerations that influence learning mathematics included in this chapter are as follows:

1. 1.0   Rate and amount of learning compared to age peers
1. 2.0   Speed of learning related to specific content
1. 3.0   Ability to retain information
1. 4.0   Need for repetition
1. 5.0   Verbal skills
1. 6.0   Ability to learn symbol systems and arbitrary associations
1. 7.0   Size of vocabulary compared with peers
1. 8.0   Ability to form relationships, concepts, and generalizations
1. 9.0   Ability to attend to salient aspects of a situation
1.10.0   Use of problem-solving strategies
1.11.0   Ability to make decisions and judgments
1.12.0   Ability to draw inferences and conclusions and to hypothesize
1.13.0   Ability in general to abstract and to cope with complexity

1.1.0 Rate and Amount of Learning Compared to Age Peers

A child's rate of learning relative to that of others of the same age is an indication of intellectual ability. Since mathematics involves abstract reasoning

3

and is expressed as a symbolic language of relationships, a child's level of intellectual functioning influences the kind and amount of mathematics he or she can learn. During the Sixties there was a great emphasis on accelerating the pace and scope of mathematics instruction. This was accomplished in two ways: first, the rate of presentation of the mathematics content was increased; and second, younger children were exposed to mathematics content that formerly had been placed at grade levels for older children. This resulted in giving first graders topics such as indirect open sentences in the form

$$7 + \square = 9 \text{ or } \square + 4 = 7,$$

and introducing children in grade eight to algebra—formerly a ninth grade subject. It became common practice to expose high school students to such college level courses as analytic geometry and calculus. Also, more content was included in the mathematics curricula at all levels. It was generally agreed that the brighter children could handle these modifications best.

On the other hand, Dunn (1973) summarized conclusions of research studies on teaching mathematics to those whose rate of learning is retarded compared to their age peers:

> The mildly retarded, as a group, tend to work up to mental age expectancy in arithmetic fundamentals, but not in arithmetic reasoning, where reading and problem-solving are involved.
>
> The mildly retarded develop quantitative concepts in the same order and stages as normal children, but considerably later. They acquire them by a combination of teaching and maturation. When drilled early to perform more advanced Piagetian-type quantitative tasks, they may appear to have the skills but will not understand the concepts, such understanding coming later after adequate intellectual growth has taken place.
>
> When compared with younger normal children of the same mental age, mildly retarded children make more careless errors in working out problems, use more primitive habits such as counting on the fingers [see discussion of Chisanbop, Wilson, 1978], get more confused by superfluous data, and have more difficulty selecting the appropriate skill to use. (p. 148)

Hood (1962), Inhelder (1968), and Woodward (1963) also found that mentally retarded youngsters progress through developmental cognitive stages in the same sequence but at a much slower rate than normals. These studies further indicated that ability for abstract reasoning is less advanced in the retarded.

1.2.0 Speed of Learning Related to Specific Content

The nature of the task may affect how fast a child learns. If the task is closely related to a strength of the child, then it will more likely be learned at a faster rate than if it were dependent upon a weakness. For example, solving a maze would be accomplished faster by one with good visual-motor coordination and/or efficient problem-solving ability, while a child with strength in verbal comprehension and expression would most likely develop a rich functioning vocabulary at a relatively earlier age than peers. Some children will quickly learn to hit a ball with a bat, while others may see cause-effect relationships faster. Some will learn the letters of the alphabet, names of colors, and symbols to represent numbers faster than their age-mates. Some will enter kindergarten already able to read; others will recognize money earlier. The rate at which children acquire content that involves verbal comprehension, perceptual organization, and numerical reasoning (Kaufman, 1975)[1,2] varies.

1.3.0 Ability to Retain Information

Memory has been described as involving retention or storage of data with no change in the way it is stored or in the way data is cued for retrieval (Guilford, 1967).

Meeker (1969) noted that memory ability was involved in various intelligence tests (Stanford-Binet, Terman & Merrill, 1960; WISC-R, Wechsler, 1974). Tasks included repeating digits, obeying simple commands, naming objects from memory, rote counting, naming the days of the week, memory for designs, copying a bead chain from memory, and reasoning. Notice that all are related in some way to mathematics ability in a numerical, temporal, or spatial sense.

Common sense suggests the importance of memory in the hierarchical nature of mathematics. Unless prerequisite ideas are remembered, new learning is difficult. For example, the configuration of the digits 0–9 must be remembered if the child is to write them on recall, and the names of the powers of 10 must be remembered in order to name place values. Memory also plays a part in processing algorithms. The following computation errors (see Figure 1.1) may have occurred because the students did not remember the sequence for performing the operations. (Other possibilities that may underlie computational errors include original learning situations, perceptually related problems, lack of developmental readiness, lack of motivation).

1.4.0 Need for Repetition

The child who sees relevant relationships within a task immediately and who has good short-term and long-term memory will need less repetition in order to remember the salient features of a situation. Appropriate drill or consolidation activities are essential, however, for children with deficits in memory ability. The observation that overlearning facilitates retention is a generally accepted tenet of learning theory.

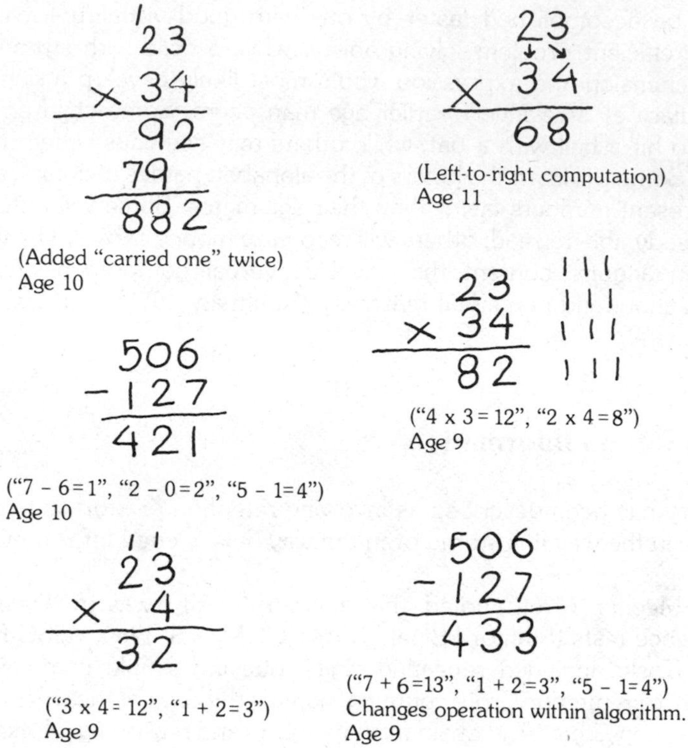

$$\begin{array}{r} \overset{1}{2}3 \\ \times\ 34 \\ \hline 92 \\ 79 \\ \hline 882 \end{array}$$

(Added "carried one" twice)
Age 10

$$\begin{array}{r} 23 \\ \times\ 34 \\ \hline 68 \end{array}$$

(Left-to-right computation)
Age 11

$$\begin{array}{r} 506 \\ -127 \\ \hline 421 \end{array}$$

("7 – 6 = 1", "2 – 0 = 2", "5 – 1 = 4")
Age 10

$$\begin{array}{r} 23 \\ \times\ 34 \\ \hline 82 \end{array}$$

("4 x 3 = 12", "2 x 4 = 8")
Age 9

$$\begin{array}{r} \overset{1}{2}\overset{1}{3} \\ \times\ \ 4 \\ \hline 32 \end{array}$$

("3 x 4 = 12", "1 + 2 = 3")
Age 9

$$\begin{array}{r} 506 \\ -127 \\ \hline 433 \end{array}$$

("7 + 6 = 13", "1 + 2 = 3", "5 – 1 = 4")
Changes operation within algorithm.
Age 9

Figure 1.1. *Computation Errors*

1.5.0 Verbal Skills

Studies of the syntax of children from two to four have shown that syntactic rules are not acquired by mere passive imitation (Bellugi & Brown, 1964) but rather involve generalizing and formulating original constructions.[3]

Verbal skills involve receptive and expressive functioning. Examples of receptive processing are comprehending words and sentences both spoken and written. Expressive verbal abilities include oral and written speech. In relation to mathematics, a student may be able to understand what he or she is reading on a flashcard—for example,

$$\boxed{\begin{array}{r} 3 \\ +2 \end{array}}$$

—and may even know the answer, but is unable to produce the label "five" or "5." The key is that the response is in the child's repertoire but cannot be expressed.

Sometimes a child is unable to comprehend the meaning of what he or she has heard or read. Since mathematics is expressed as a symbolic language, a child with a receptive problem will encounter difficulty understanding mathematical as well as general language symbols. On the other hand, it should not be assumed that a correct imitative response indicates that the child has syntactic or semantic knowledge—it may just be a rote memory response.

Siegel and Broen (1976), in a chapter on language assessment, pointed out that morphemes such as inflectional endings indicate tense, possessives, and plurals (numerosity). Young children show incomplete mastery of these forms—"*I want two toy, I gived my sister a cookie, This is John mother.*" Other quantitative ideas such as comparatives and superlatives are also expressed by inflection.

The teacher must be aware of mathematical relationships that are embedded in our general language. Mathematical language may be seen as a core of precision within symbolic communication. An equation yields maximum meaning with a minimum of components. Items designed to assess quantitative, spatial, and temporal aspects of general language will be discussed in Chapters 2, 6 and 8.

1.6.0 Ability to Learn Arbitrary Associations and Symbol Systems

Mathematics is communicated by means of various symbol systems. "In order to communicate thoughts . . . there must be a conventional system of signs or symbols which when used by some persons, are understood by other persons receiving them" (Gelb, 1963, p. 1). These symbols are arbitrarily associated with ideas they represent. As graphic systems evolved over time, they became increasingly more systematized. However, in spite of increased use of systematization, acquisition of language in graphic form is very difficult for some children.

Clark and Woodcock (1976, p. 552) defined a graphic system as "a system comprised of a rule-governed set of visible marks that provide a means of linguistic intercommunication." Mathematics incorporates logographs (for example, digit, addition sign, square root sign, and so on), in which the symbol unit represents one or more words which in turn represent a complete thought. Syllables stand for two or more phonemes or unit speech sounds. Alphabetic systems use a few symbols in combinations to form thousands of different words. In some languages there is a one-to-one relationship between grapheme (letter) and phoneme (sound). However, in written English there are many irregularities in phoneme–grapheme correspondence.

Although logographic systems do not have the efficiency of systems based upon syllables or alphabets, they do have some advantages (Clark & Woodcock, 1976):

logographs represent word units or ideas, and are probably easier to learn and use initially than systems that represent units smaller than words. . . . Some logographic systems and many symbols within systems can be understood in almost any language. . . . The abstract logographs of a mathematical language . . . know few language boundaries. $1 + 1 = 2$ can be read by speakers of many languages. A concrete pictograph of a cow represents that animal in any language whereas the alphabetically written word cow is decipherable only to readers of English. . . . Pictographs are often referred to as *rebuses* [see Figure 1.2]. Rebuses

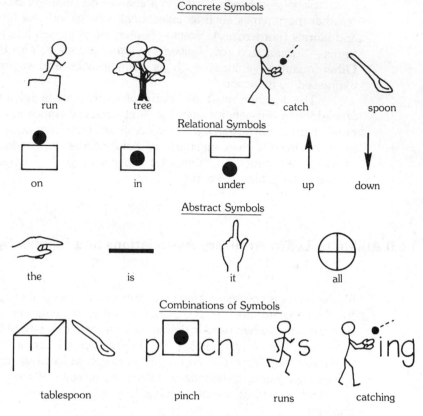

Figure 1.2. *The Three Types of Rebus Symbols, and Word
Combinations Utilizing Rebuses with Letters*

From Clark, C.R., & Woodcock, R.W., Graphic systems of communication. In L.L. Lloyd (Ed.), *Communication assessment and intervention strategies* © 1976 by University Park Press, Baltimore. Reprinted
by permission.

are used widely as international symbols for equipment controls and road signs.
. . . Rebus symbols may be classified into three types: concrete, relational, or
abstract. (pp. 582–83)

Another symbol system is that of Bliss (1965), which is based on symbolic
logic, semantics, and ethics (see Figure 1.3). "Bliss symbols are both ideographic and pictographic. By ideographic it is meant that a Bliss symbol
represents an idea or a concept that may be expressed by the child.
Pictographic implies that the symbol is also 'picture-like' " (Vanderheiden &
Harris-Vanderheiden, 1976, p. 636).

A child's ability to learn mathematics symbols is often confused with
perception. Perception is not a symbolic function, because percepts rely upon
the here and now. Take a honeydew melon, for example. You can see it, feel it,
smell it, taste it, feel if it's cold or warm, pick it up, and hear its thud if you drop
it. The melon is a given; you do not have to imagine its existence—it is real. Not

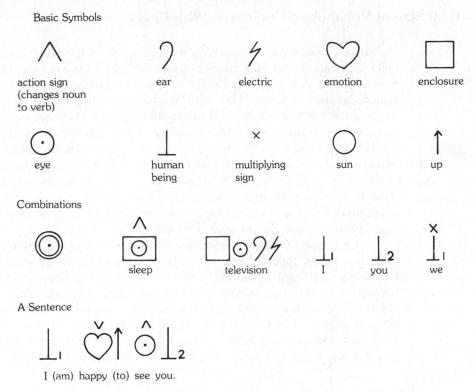

Figure 1.3. *Selected Examples of Bliss Symbols*

From Clark, C. R., & Woodcock, R. W., Graphic systems of communication.
In L. L. Lloyd (Ed.), *Communication assessment and intervention strategies*
© 1976 by University Park Press, Baltimore. Reprinted by permission.

so with number. Number is conceptual. It is an abstraction. Arieti (1976, p. 98) suggested that "the general process of cognition . . . is based on two fundamental characteristics: progressive abstraction and progressive symbolization . . . and that abstraction leads to symbolization." The use of concrete representations of number facilitates this progressive abstraction and its relation to symbols.

However, it should be remembered that a mathematics symbol is a primary symbol in the sense that it directly represents an abstraction. The digit "2" stands for the abstraction "twoness"; the plus sign "+" signifies an arithmetic operation; the subset sign "C" implies an inclusion relationship; other signs—for example, the square root sign "$\sqrt{}$" or pi "π"—all have precise referents. These referents do not have concrete counterparts, objects in the environment whose attributes can be perceived with the senses. For example, attributes of all girls will be pretty much the same in a physiological sense. This is not so with equivalent sets that represent a number idea such as "three." Thus, there is a high degree of abstraction embedded in mathematical symbols, and a degree of ability in abstract reasoning is necessary in order to comprehend a mathematical symbol system.

9

1.7.0 Size of Vocabulary Compared With Peers

Vocabulary assessment involves more than a tally of the number of words a child understands and uses. Since a given word may have a number of different meanings and nuances, it is helpful to obtain a child's depth of vocabulary knowledge. For example, "He's cool" might mean either that he is unfriendly, needs a sweater to warm up, or is a desirable person. "The more meanings and nuances the child can pull from a vocabulary item, the richer his command of the language" (Siegel & Broen, 1976, p. 80). It has been found that hearing-impaired children and retarded children develop an incomplete understanding of the meaning of some words. Mathematically related words are particularly difficult. For example, you might teach the word "dog" by showing the child a picture of a dog or having a dog present for the child to pet and say "This is a dog." However, you do not say, "This is a *more*."

Comparing a child's vocabulary with that of his peers may give you an indication of the child's ability to conceptualize and is a measure of acquired knowledge (see Bannatyne, 1971). When one observes size and/or comprehension of a child's vocabulary, the impression obtained from a single situation (such as only in a school or classroom setting) may not be representative of other situations. Since vocabulary skills are often taken as part of general intellectual skills, it is important to assess a child's vocabulary in the broadest possible way and to be aware of cultural and experiential limitations.

1.8.0 Ability to Form Relationships, Concepts and Generalizations

Examples of basic lower-level relationships in mathematics are sequencing and matching elements of a set to elements of another set, equal in number on a one-for-one match. The ability to sequence involves attending in succession and then arranging objects or events into a sequence. Out of the ability to sequence emerges the notion of one-to-one correspondence.

One-to-one correspondence is one of the most basic relationships in mathematics. Unless a child understands the one-to-one relationship, he or she does not comprehend identity or the relation of equality. One-to-one correspondence underlies the following mathematical ideas: many-to-one, equality, and greater than–less than.

When the one-to-one relationship is generalized to more than two sets, the generalization of equivalent sets is developed. This is an example of a lower-level generalization. Next, the child may state that several sets have the same number of elements. The child who can identify the cardinal number property of equivalent sets is displaying a specific concept as "threeness."

Examples of higher-order relationships are binary operations, including addition, multiplication, subtraction and division; and unary operations, such as squaring or cubing a number and finding the root of a number. When a child applies an arithmetic operation to numbers, he or she is generalizing at the rule or principle level—for example, putting two concepts together into some sort of relationship, as in

"two plus three equals five" (2 + 3 = 5).

Figure 1.4 shows how arbitrary associations, relationships, concepts and generalizations are related. (This hierarchy—the developmental mathematics curriculum—forms the structure for Section II of this book.)

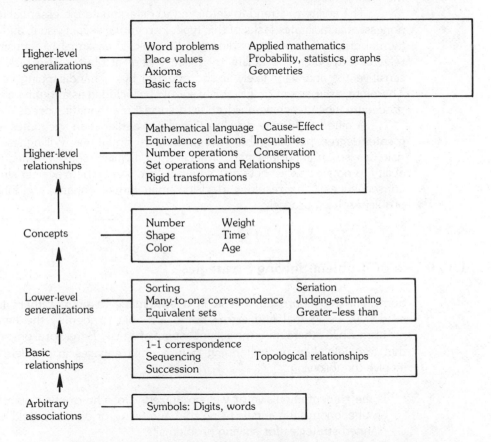

Figure 1.4. *Development of Arbitrary Associations, Relationships, Concepts, and Generalizations—Structure for Understanding the Psychological Nature of a Mathematics Curriculum*

1.9.0 Ability to Attend to Salient Aspects of a Situation

Ability to notice the important and most relevant aspects of a situation is extremely important in mathematics learning. This cognitive function underlies abstracting and classifying processes. In order for a child to abstract the number property of equivalent sets, it is necessary to disregard extraneous cues such as color, size, shape, texture, or position of objects in space. Focus must be on the number of elements, that is, on the fact that all of the equivalent sets being considered have the same number of elements. On the other hand, if the color red is the salient feature under consideration, then all other attributes except "redness" must be disregarded. Thus, the child has to make a conscious choice as to what attribute or aspects of a situation are essential to the task at hand.

The ability to attend to detail and to differentiate the essential from the nonessential facilitates tasks of this type. Perceptual skill in visual, auditory, or tactual discrimination, depending on the task, is essential to focusing on a relevant attribute. The ability to notice salient attributes of a task is most sensitive to, and very likely to be affected by, a handicapping condition. Therefore, assessment of this ability should be included as a routine diagnostic procedure when teaching children with special mathematics needs.

A gifted child may develop this ability earlier than age-mates and to a greater degree, providing there are no sensory limitations. Willingness to take risks is also an important part of successful problem solving, reflected in an ability to notice risk aspects of a situation and weigh them against alternative consequences. The resulting decision may either enhance or inhibit the problem-solving process.

1.10.0 Use of Problem-Solving Strategies

Some children approach a task in a systematic manner. Others flounder randomly, using trial and error. Some are able to identify the heart of a problem while others miss the salient aspect of a task. Behavioral observations that are helpful in understanding how a child engages in problem solving involve the following:

- Is the student impulsive or well-organized? Does he or she skip around in an uneconomical manner, using trial and error, or does the child use well-planned strategies for solving problems?[4,5]
- Does the student become discouraged quickly and give up on a task, or does he show persistence in working toward task completion? Does he need encouragement to continue, or is he independent and self-sufficient?
- Does the student modify his or her pace while performing a lengthy task? Does he continue at one rate throughout, or does he alter his performance as need arises? An appropriate increase in speed, especially during a timed task, implies ability to adjust to the demands of the particular task. A decrease in speed of performing a task accompanied by correct responses may be an indication of fatigue. Observe whether the child can correctly solve a problem that was missed due to time constraints in an untimed situation.

Behavioral observations are an important prerequisite before a teacher can infer what a child's performances mean. Objective, nonevaluative observation should occur first. However, the teacher should be sensitive to the possibility that the behavior observed may lead to a variety of inferences or hypotheses explaining the student's actions. For example, it is important to know that the potential is present for well-planned problem solving, but emotional disturbances are interfering.

A widely used behavioral observation that is related to problem solving is the IQ score. The IQ score is only an indication of the level of cognitive functioning. Many causes can interfere with obtained IQ—for example, the rapport established between the psychologist and the student; the student may have not slept well the night before; the student might not have a desire to do well; or on the other hand, the child may be on his or her best behavior and perform better than is usually observed in school settings. It is extremely important for a teacher to observe the way a child proceeds to obtain a score. The quality of the test taker's behavior is as important as the score.

1.11.0 Ability to Make Decisions and Judgments

Previous experience or familiarity with a situation aids one in making decisions. Making decisions and judgments involves recognizing salient aspects of a situation, appreciating important information given, being aware of incompleteness of information, being able to determine what is missing, and abstracting essential from nonessential details. The child must be able to evaluate relationships embedded in a situation and be able to make choices among alternatives. Also involved is the ability to inhibit irrelevant alternatives so that these do not interfere with the decision-making process.

1.12.0 Ability to Draw Inferences and Conclusions, and to Hypothesize

The ability to infer and hypothesize is a component of creative problem solving. In order to make inferences, an individual needs to generate possible alternatives and deal with future ideas. Judgments must be made according to a set of criteria. These processes involve forming relationships and concepts and then generalizing to select a solution to the problem at hand.

1.13.0 Ability to Abstract and Cope With Complexity

Evidence of the degree of abstractness and complexity with which an individual can cope includes classifying objects or ideas, finding logical relationships or analogies, performing simple operations of logical deductions, using similes and metaphors, and solving constructive tasks (as in the block design subtest of the WISC-R). Children developing at a slower cognitive rate do not handle abstract and complex situations well. Therefore, ideas must be concretized for these learners. Conditions that may be expected to affect a child's ability to abstract and to comprehend complex ideas include the following:

* Inability to combine separate elements of information into a unified organization
* Disturbances of short- or long-term memory
* Inability to inhibit impulsive attempts to find an answer to a question
* Inability to make preliminary investigations of the basic conditions of a task in order to identify the problem to be solved
* Inability to generate hypotheses and to choose, from many possible alternatives, strategies for converting a complex problem into a series of manageable tasks

(Suggestions for reducing the abstract and complex nature of tasks are presented in Chapters 5 through 12 in Section II. Modifications of mathematics instruction for slow-learning as well as rapid-learning students are included in Chapters 5 through 9.)

Summary Thought Questions and Activities

I. You should now be aware of cognitive factors that may affect mathematics learning.
    A. Which of these factors affected your own mathematics learning?
    B. Which factors enhanced your own educational experiences in mathematics?
    C. Which inhibited your success in mathematical reasoning?
II. List all ideas and vocabulary in this chapter that are new to you. Compare your list with the chapter contents. Identify ideas you did not remember but that you consider essential in teaching an exceptional child.
III. Interview a child's teacher regarding his or her observations of generic cognitive factors discussed in this chapter that influence learning mathematics.
IV. Describe how you would employ generic factors that affect mathematics learning to select appropriate mathematics topics for a student. For example, how could a generic factor affect selection of addition computations ?

Suggested Readings

Bijou, S. W. Environment and intelligence: A behavioral analysis. In R. Cancro (Ed.), *Intelligence· Genetic and environmental influences.* New York: Grune and Stratton, 1971.

14

Bracht, G. H. Experimental factors related to aptitude—treatment interaction. *Review of Educational Research*, 1970, *49*, 627–745.

Engleman, S. The effectiveness of direct verbal instruction in IQ performance and achievement in reading and arithmetic. In J. Hellmuth (Ed.), *Disadvantaged child Volume 3: Contemporary education*. New York: Brunner/Mazel, 1970.

Fliegler, L. A., & Bish, C. E. Summary of research on the academically talented student. *Review of Educational Research*, 1959, *29*, 408–450.

Fodor, J. D. *Semantics: Theories of meaning in generative grammar*. Hassocks, Sussex: Harvester Press, 1977.

Frish, S. A., & Schumaker, J. B. Training generalized receptive prepositions in retarded children. *Journal of Applied Behavior Analysis*, 1974, 7, 611–621.

Gowan, J. C. Imagery, incubation, and creativity. *Journal of Mental Imagery*, 1978, *2*, 23–32.

Hadamard, J. *An essay on the psychology of invention in the mathematical field*. Princeton: Princeton University Press, 1949.

Kamii, C. K. Evaluation of learning in pre-school education: Socio-emotional, perceptual-motor, cognitive development. In B. S. Bloom, J. T. Hastings, & G. F. Madaus (Eds.) *Handbook of formative and summative evaluation of student learning*. New York: McGraw-Hill, 1971.

Weikart, D. P., Rogers, L., Adcock, C., & McClelland, D. *The cognitively oriented curriculum*. Washington, D. C.: National Association for the Education of Young Children, 1971.

Wilson, J. W. Chisanbop: Sales hype or substance? *Reflections: September 1978, 26*, 1. (Atlanta: Official Publication of The Georgia Council of Teachers of Mathematics.)

References

Bannatyne, A. *Language, reading and learning disabilities*. Springfield, Ill.: Charles C Thomas, 1971.

Bellugi, U., & Brown, R. (Eds.) The acquisition of language, *Monographs of the Society for Research in Child Development*, 1964, *29* 1–192.

Bliss, C. K. *Semantography*. Sidney, Australia: Semantography Publications, 1965.

Clark, C. R., & Woodcock, R. W. Graphic systems of communication. In L. L. Lloyd (Ed.), *Communication Assessment and Intervention Strategies*. Baltimore: University Park Press, 1976.

Dunn, M. *Exceptional children in the schools—special education in transition* (2nd ed.). New York: Holt Rinehart and Winston, Inc., 1973.

Gelb, I. J. *A Study of writing* (rev. 2nd ed.). Chicago: University of Chicago Press, 1963.

Gruber, H. E., & Voneche, J. J. *The essential Piaget*. New York: Basic Books, 1977.

Guilford, J. P. *The nature of human intelligence*. New York: McGraw-Hill, 1967.

Harris, L. J. Neurophysiological factors in the development of spatial skills. In J. Eliot & N. J. Salkind (Eds.), *Children's spatial development*. Springfield, Ill.: Charles C Thomas, 1975.

Hood, H. B. An experimental study of Piaget's theory of the development of number in children. *British Journal of Psychology*, 1962, *53*, 273–286.

Inhelder, B. *The diagnosis of reasoning in the mentally retarded* (2nd ed); W. B. Stevens and others, trans.). New York: Chandler Publishing, 1968.

Kaufman, A. S. Factor analysis of the WISC-R at 11 age levels between 6½ and 16½ years. *Journal of Consulting and Clinical Psychology*, 1975, *43*, 135–147.

Krutetskii, V. A. *The psychology of mathematical abilities in school children* (J. Kilpatrick & I. Wirszup, eds., and J. Teller, trans.). Chicago: University of Chicago Press, 1976.

Luria, A. R. *Speech and intellect in child development.* Moscow: Izd. Akad. Komm. Vospit., 1928.

Luria, A. R. The neuropsychological study of brain lesions and restoration of damaged brain functions. In M. Cole & I. Maltzman, (Eds.), *A handbook of contemporary Soviet psychology.* New York: Basic Books, 1969.

Luria, A. R. *The working brain, An introduction to neuropsychology.* New York: Basic Books, Inc., 1973.

Meeker, M. N. *The structure of intellect.* Columbus: Charles E. Merrill, 1969.

Siegel, G. M., & Broen, P. A. Language assessment. In L. L. Lloyd (Ed.), *Communication assessment and intervention strategies.* Baltimore: University Park Press, 1976.

Terman, L. M., & Merrill, M. A. *Stanford-Binet intelligence scale.* Boston: Houghton-Mifflin, 1960.

Vanderheiden, G. C., & Harris-Vanderheiden, D. Communication techniques and aids for the nonvocal severely handicapped. In L. L. Lloyd (Ed.), *Communication assessment and intervention strategies.* Baltimore: University Park Press, 1976.

Vygotskii, L. S. Defect and overcompensation. *Pedagogical Encyclopaedia*, 1928, *11*, 391–392.

Vygotskii, L. S. The psychology of schizophrenia. In *Contemporary problems of schizophrenia.* Moscow: Medgiz, 1933.

Vygotskii, L. S. The problem of mental retardation. In *Selected psychological investigations.* Moscow: Izd. Akad, Pedag. Nauk RSFSR, 1956.

Vygotskii, L. S. Thought and language. Cambridge, Mass.: MIT Press, 1962.

Wechsler, D. *Manual for the Wechsler intelligence scale for children—revised.* New York: Psychological Corporation, 1974.

Woodward, M. The application of Piaget's theory to research in mental deficiency. In N. R. Ellis (Ed.), *Handbook of mental deficiency.* New York: McGraw-Hill, 1963.

Endnotes

1. Harris (1975) attributed the content-specific phenomenon to the fact that the two hemispheres of the brain have very different functions:

    The left hemisphere—that which controls the right side of the body—is intimately involved in the analysis of language and symbolic material, while the right hemisphere subserves nonlinguistic, visuo-spatial functions. . . . The respective specialization of the left and right cerebral hemisphere for linguistic and nonlinguistic targets in the realm of vision are matched in audition. Speech sounds, such as spoken digits and other words, are heard more accurately by the left hemisphere, while nonspeech sounds, such as coughing, laughing, crying, or melodic patterns, are heard more accurately by the right hemisphere. (p. 6)

2. Luria (1973) studied the "learning-related-to-content" situation from a neuropsychological view. Through neuropsychological analysis he demonstrated profound differences in the character of disturbance of problem solving by patients with lesions of different parts of the brain.

3. Luria (1928) pointed out the problem of speech and its role in the formation of mental processes. Complex mental processes are formed with the intimate participation of language during the child's association with adults. It is for this reason that study of verbal skills is important in ascertaining how social relationships are developed in the child, how he or she masters language, how with the aid of speech the child masters the experience of prior generations, and finally how speech aids the formation of higher conscious mental activity. Vygotskii (1928, 1933, 1956) showed that "higher mental processes (active attention, voluntary recall and abstract thinking) develop as a result of the child's interaction with adults and his experimentation with objects and speech" (Luria, 1969, p. 282). Luria (1969) further described how Vygotskii emphasized the role of speech in the organization of all human behavior:

Vygotskii . . . carried out a series of studies on the basic mental processes (active attention, voluntary memory, volitional behavior). His findings showed that the meaning of a word, which is the basic instrument of abstraction and generalization, changes during the course of the child's development. . . . An experimental approach was used in the analysis of how the forms of reality are experienced at successive stages of development, how these forms are disturbed in pathological states of the brain, how relationships among basic psychological processes (perception, recall, and abstract-generalizing speech) change at the various stages of ontogenesis, and how they change in the various forms of pathological alterations of mental activity.   (p. 282)

Vygotskii believed that "for a long time the word is to the child a property, rather than the symbol, of the object; that the child grasps the external structure word-object earlier than the inner symbolic structure" and that "ontogenetically thought and speech develop along separate lines and that at a certain point these lines meet" (1962, pp. 19–20).

Piaget's description of the development of language is presented by Gruber and Voneche (1977, pp. 504–506):

In the normal child, language appears at about the same time as the other forms of semiotic [symbolic] thought. In the deaf-mute, on the other hand, articulate language does not appear until well after deferred imitation, symbolic play, and the mental image. . . . Deaf mutes . . . elaborate a gestural language . . . [that] is both social and based on imitative signifiers that occur . . . in deferred imitation, in symbolic play, and the image. . . . Articulate language makes its appearance . . . at the end of the sensorimotor period with . . . one-word sentences . . . [that] express in turn desires, emotions, or observations. From the end of the second year, two-word sentences appear, then short complete sentences without conjugation or declension, and next a gradual acquisition of grammatical structures.

4. Krutetskii (1976, p. 183) described characteristics of mental activity in solving mathematical problems:

Apparently, three links or stages can always be traced in the solution of any problem (from the elementary to the very complicated). The solution of any problem seems to begin with the acquisition of initial facts, initial information about the problem, with thorough reflection, attempts to understand, and mastery. Then comes the solution proper, as a stage of processing or transforming the facts acquired for the purpose of attaining the desired result. And finally, both the process and the result of the solution always leave some trace in the memory, somehow enriching a person's experience.

5. Luria (1973, pp. 336–340) pointed out that:

different problems have different structures, and it is possible to arrange their structures in the order of increasing complexity. For instance, the simplest problems (Jack has four apples, Jill has three apples. How many apples have they together?) have a simple algorithm of solution and no special searching is required. More complex forms of this problem (Jack had four apples, Jill had two apples more. How many apples had they together?) require the performance of an intermediate operation, which is not expressed in the words in the problem, so that the solution of the task breaks up into two steps: $a + x = y$: $x = (a + b)$; $(2a + b) = y$. The solution of problems with complex algorithms, requiring programmes consisting of a series of successive components, is much more difficult. . . . These include recoding of the conditions and the introduction of new component elements. (There were eighteen books on two shelves; there were twice as many books on one shelf as on the other. How many books were there on each shelf? To solve this problem, in addition to the original condition "on two shelves" the additional items "three parts" must be introduced and the number of books on each of these auxiliary "parts" must first be calculated.) Finally, perhaps the most difficult problems are those described as "conflicting" and in which the correct method of solution involves the inhibition of the impulsive direct method. (A candle is 15 cm long; the shadow from the candle is 45 cm longer; how many times is the shadow longer than the candle?) The tendency here is to perform the direct operation $45 \div 15 = 3$; this must be inhibited and replaced by the more complex programme: $15 + 45 = 60$; $60 \div 15 = 4$. Lesions of the left temporal region, disturbing audio-verbal memory, give rise to difficulty in retention and are accompanied by inability to involve the necessary intermediate speech components. . . . For this reason the solution, even of relatively simple problems, is severly impaired in patients of this group. The process can be facilitated to some extent if the problem is presented in writing, but even in these cases the need for intermediate speech components, used as elements for the solution of problems, severely impairs the entire discursive process. The difficulties in the solution of problems encountered by patients with lesions of the systems of the left parieto-occipital region . . . causes gross impairment of simultaneous (spatial) syntheses, and this is manifested both in the direct, concrete behavior and also in the symbolic sphere. As a result of such a disturbance it becomes impossible to operate either with logical-grammatical systems or with systems of numerical operations, so that the normal solution of complex problems is prevented. The general meaning of the problems is often relatively intact in these patients; as a rule they never lose sight of the final question of the problem and they make active attempts to find a method of solution. However, the fact that they cannot understand complex logical-grammatical structures or perform any [other] than the simplest arithmetical operations presents an insuperable obstacle to their problem solving. Even such elements of the conditions as "A took so many apples from B," or "A had twice as many apples as B," or again, "A had two apples more (or less) than B," not to mention more complex logical-grammatical structures, are quite beyond the capacity of such patients. . . . The disturbance of the process of problem solving differs completely . . . in patients with lesions of the *frontal zones* . . . when given a written problem, they do not perceive it as a problem, or in other words, as a system of mutually subordinated elements of the condition which must lead to the solution of the problem. If such patients are instructed to repeat the conditions of the problem, they may succeed in reproducing some of the component elements of the condition, they may be unable to repeat the problem at all, or instead of repeating the problem they may repeat only one of its elements. As an example, let us consider the problem "There were eighteen books on two shelves, but they were not equally divided; there were twice as many books on one shelf as on the other. How many books were there on each shelf?" The patients of this group will repeat the problem as follows: "There were eighteen books on two shelves. And on the second shelf there were eighteen books. . . ." The patients may repeat the same problem but in a different way: "There were eighteen books on two shelves; there were twice as many on one shelf as on the other. How many books were there on both shelves?" In this case they do not observe that . . . they are in fact repeating the first part of the conditions. The patients of this group, as a rule, are unaware of their mistake, and if the problem is presented to them again, they simply repeat their previous mistakes. These facts clearly show that in a patient with a marked frontal syndrome the basic condition, that of the existence of the problem itself is lacking, so that there can be no intention to "solve" it. The second defect characteristic of patients with a frontal lobe lesion, intimately connected with the first, is that these patients make no attempt at a preliminary investigation of the conditions of the problem, and as a result, without any preliminary analysis of the conditions and identification of their components, they immediately begin to seek solutions impulsively, usually by combining the numbers specified by the conditions and performing a series of fragmentary observations, totally unconnected with the context of the problems . . . . A typical example of these fragmentary operations, which replace the true solution of the problem, is given by the course of the answers to the problem of the two shelves with eighteen books: "There were eighteen books on two shelves . . . there were twice as many on the second . . . thirty-six . . . and there were two shelves . . . 18 + 36 = 54! . . ." This type of "solution" of a problem is typical of patients with a marked frontal syndrome, and it clearly demonstrates the disintegration of intellectual activity as a whole in these patients, so that they cannot solve such problems despite the fact that their understanding of logical-grammatical structures and of arithmetical operations is intact. In such cases, patients with a frontal lobe lesion characteristically do not compare their answer with the original conditions of the problem and they are unaware of the meaninglessness of their solution. The correct solution to such "conflicting" problems is clearly quite beyond the capacity of these patients. For example, one such patient, having read out the problem—"A candle is 15 cm long; the shadow from the candle is 45 cm longer. How many times is the shadow longer than the candle?"— immediately answers "three times of course!" and even after leading questions he fails to notice anything wrong about his answer.

From *The working brain: An introduction to neuropsychology* by A. R. Luria, pp. 336–340, © Penguin Books Ltd. 1973, translation © Penguin Books Ltd., 1973, Basic Books, Inc., Publishers, New York. Reprinted by permission.

Psychomotor Factors That Influence Learning Mathematics

2

Generic factors that affect mathematics learning include inefficiency in the use of psychomotor abilities, including those abilities needed in searching as well as in producing spoken or written responses. The following influences are discussed:

2.0.0 Psychomotor Abilities or Motor Effects of Mental Processes
2.1.0 Disorders of Visual Perception
2.1.1 Poor Visual Discrimination
2.1.2 Visual Figure-Ground Distractibility
2.1.3 Form–Constancy Problems
2.1.4 Deficit in Visual-Sequential Memory
2.1.5 Difficulties in Spatial Relationships
2.2.0 Disorders of Auditory Perception
2.2.1 Poor Auditory Discrimination
2.2.2 Auditory Figure-Ground Distractibility
2.2.3 Poor Auditory Analysis
2.2.4 Sound-Blending Difficulty
2.2.5 Problems in Auditory Sequential Memory
2.3.0 Rules of General Language and Mathematics
2.3.1 Phonological Rules
2.3.2 Morphological Rules
2.3.3 Syntactical Rules
2.3.4 Semantic Rules

2.0.0 Psychomotor Abilities or Motor Effects of Mental Processes

Strauss and Lehtinen (1947) hypothesized that children with impaired psychomotor abilities have difficulty with mathematics because they lack ability to spontaneously discover significant relationships of the number system. This was due to their disordered "visual-spatial organization." Strauss and Lehtinen emphasized the importance of being able to visually perceive schemes of spatial organization in forming number concepts. Hebb (1949) also emphasized the role of visual perception in developing mathematical ideas—in particular, recognizing and discriminating forms and shapes. Hebb pointed out that controlled eye movement is essential to perceptual integration.

Children with defects resulting from visuomotor (movement under visual control; Abercrombie, 1963) disorder have difficulty recognizing shapes, matching, and discriminating forms, and attending to salient attributes of a situation. These skills of spatial ability underlie the development of number concepts in young children. To a young child, "fourness" takes up more space than "twoness." Number is a spatial idea; number is also a temporal idea. When one enumerates objects in a set by matching, in order, a number name to each element of a set, the elements are attended to in sequenced time. We cannot count simultaneously. The ability to search—that is, to scan in a controlled manner—is therefore essential to developing concepts of number, geometric shapes, and spatial relationships underlying topological ideas (order, proximity, separateness, enclosure).

Psychomotor learning appears to develop in three stages: the reception of experience (Stage 1), the central organization of received experience (Stage 2), and the expression of what has been learned (Stage 3). Interference in development of any of these stages will affect acquisition of mathematical ideas. Koupernik, MacKeith, and Williams (1975) summarized these three stages:

> In the beginning of learning, experience comes to us all primarily through the channels of our senses. We learn about our world from what we see, from what we hear, from the experiences gained through touching and feeling and handling objects, and, later, through all the various experiences gained through movement. If any of these means of gaining experience are faulty, or if the opportunity for the child to gain experience in these ways is lacking, the possibility for development will be impaired to some extent. . . . Central organization is the process by which input gained through the sensory channels and experience . . . is structured and organized to become meaningful. . . . The capacity for central organization involves the formation of associations and the understanding of relationships, the ability to recall previous experiences, to form concepts, to integrate information gained through one sense modality with what has been gained through another, and to relate new experience. . . . This ability depends upon the integrity of the brain and its level of maturation and . . . can be judged through the third stage, described as output.   (p. 126)

Output is the child's ability to produce some kind of response to demonstrate that he has learned. Examples include spoken or written responses. Appropriate output is dependent upon psychomotor abilities of Stage 2—the central organization of received experience. This is often referred to as "perception." Perception, the meaningful organization of received experience, can inhibit acquisition of mathematical relationships and concepts "if the child suffers dysfunction in any of the following areas of stimulus—vision, hearing, fine and gross motor coordination, and language" (Koupernik et al., 1975, p. 127). The ability of the brain to select salient stimuli and to relegate the less important to the background of our consciousness relates to ability to abstract number. For example, given a set of multicolored objects grouped in sets of three, if the child is to abstract the cardinal number property *three* from each set, she must attend to number and not to color, shape, size, texture, position, or function of elements in a set.

2.1.0 Disorders of Visual Perception

Visual perception involves the young child's understanding his world through visual experience—through what he sees. Visuoperceptual disorders underlie difficulties in spatial orientation, recognizing position, discriminating figure from ground, and distinguishing near–far relationships.

Clumsiness often accompanies visual-perceptual difficulties. Severity of this clumsiness varies with respect to the child's chronological age and general level of intelligence. This causes difficulty in arithmetic computation involving writing of algorithms. The child with poor fine motor coordination skills in addition to perceptual problems has trouble aligning digits according to place value and processing the various algorithms. (Suggestions for instructional strategies are presented in Chapter 10.)

Compton and Bigge (1976) listed behaviors children display consistently that interfere with efficient visual processing:

1. *Poor Visual Discrimination*—the inability to perceive the difference between two similar visual symbols.
2. *Visual Figure-Ground Distractibility*—the inability to screen out irrelevant visual stimuli.
3. *Form-Constancy Problems*—the inability to perceive similarity in meaning in visual stimuli with slightly different forms; a generalization process.
4. *Deficit in Visual-Sequential Memory*—the inability to process or recall a series of visual stimuli in sequential order.
5. *Difficulties in Spatial Relationships and Directionality*—the inability to organize visual stimuli in space; difficulty with left-to-right or top-to-bottom orientation. (p. 192)

Examples of how these behaviors represent psychomotor factors that affect mathematics learning follow.

2.1.1 Poor Visual Discrimination

1. Confuses operational signs: + and ×, or − and ÷.
2. Confuses set notation:

3. Confuses digits 6 and 9, 4 and 7.
4. Confuses geometric shapes:

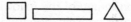

2.1.2 Visual Figure–Ground Distractibility

1. Unable to disregard irrelevant attributes, as in the Piagetian task dealing with conservation of mass. Here the child is shown two equivalent balls of

clay. When one is rolled into a longer thinner shape, the nonconserving child believes that a change in mass has also occurred.

2. Unable to attend to number property of equivalent sets and simultaneously disregard color, size, shape, texture, or function of objects in the sets.

3. Unable to notice topological equivalents such as simple closed curves; for example, ◯ and ⟳ are topological equivalents whereas ◯ and 8 and ⟨ are not.

4. Difficulty in solving mazes.

2.1.3 Form-Constancy Problems

1. Does not recognize digits 0, 1, 2, 3, 4, 5, 6, 7, 8, 9 if they are presented with variation in size, color, or position in room.

2. Does not recognize that there are an infinite number of names for a number:

$$4 = 2 + 2 = 8 \div 2 = 100 - 96 = 1 \times 4 = 32/8 \text{ etc.}$$

3. Understands congruence of geometric figures but not similarity due to change in size.

2.1.4 Deficit in Visual-Sequential Memory

1. Reverses digits in a numeral: Reads 473 as 437.
2. Unable to write digits in order from 0–9.
3. Unable to label geometric figures in conventional sequence:

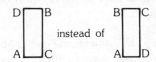

4. Unable to record ordered elements of a set representing the cross product (×) of two sets:

$$A = \{ 1, 2, 3 \}$$
$$B = \{ 4, 5 \}$$
$$A \times B = \{ (1, 4), (1, 5), (2, 4), (2, 5), (3, 4), (3, 5) \}$$

2.1.5 Difficulties in Spatial Relationships

1. Confuses decimal notation: 7.34 and .734.
2. Confuses multiplication and decimal notation: 3·5 and 3.5.
3. Writes ℮ for 6 and Ɛ for 3.

4. Difficulty comprehending topological notions: *enclosure* (the dot is inside the circle ⊙), *order* (the triangle is third in line to the right ☐ ○ △), *proximity* (draw a ball next to the tree o🌳), and *separateness* (an open curve ⌇).

5. Difficulty seeing rotations and projections.

6. Is unable to write algorithms in an orderly fashion:

$$\begin{array}{r} 32 \\ \times 7 \\ \hline 14 \\ 210 \\ \hline 224 \end{array}$$

Child may write
algorithm as:

$$\begin{array}{r} 3\,{}^{2} \\ \times 7 \\ \hline 2114 \end{array}$$

or

$$\begin{array}{r} {}^{\prime} \\ 32 \\ \times 7 \\ \hline 4 \\ 22 \\ \hline 26 \end{array}$$

2.2.0 Disorders of Auditory Perception

Compton and Bigge (1976) defined auditory perception as:

the organization and interpretation of auditory stimuli—both speech sounds and environmental sounds—in the brain. It is the internal process of organizing things we hear and interpreting them on the basis of previous experience. It is discriminating, categorizing, and giving meaning to sounds. The process takes place in the brain and not in the ears. (p. 191)

Since speech has a temporal underpinning involving rhythm, as does number, disability in auditory perception is a basic factor that affects mathematics learning.

Some behaviors that depend upon auditory skills were presented by Compton and Bigge:

1. *Poor Auditory Discrimination*—the inability to perceive the difference between two similar sounds; that is, the misperception of one word for another.

2. *Auditory Figure-Ground Distractibility*—the inability to screen out irrelevant auditory stimuli and focus upon the primary auditory message.

3. *Poor Auditory Analysis*—the inability to break down a spoken word into individual sounds and analyze their number and order.

4. *Sound Blending Difficulty*—the inability to perceive a sequence of individual sounds and then blend them into a word; . . . also called *auditory synthesis*.

5. *Problems in Auditory Sequential Memory*—the inability to recall a series of auditory events *in order*; child cannot repeat a telephone number or a sentence, retell a story, or name the days of the week in order. (p. 191)

Examples of how these factors affect mathematics learning are as follows:

2.2.1 Poor Auditory Discrimination

Child draws *hair* instead of a *square*. Child gets *in* instead of *on* a box.

2.2.2 Auditory Figure-Ground Distractibility

Child has difficulty hearing oral drill on basic addition facts due to noise of flashcards being moved. No one else is distracted. Child cannot count beats on a drum arranged in rhythm.

2.2.3 Poor Auditory Analysis

Child has difficulty when asked to express plurals; for example, "I have three sister" for "I have three sisters."

2.2.4 Sound Blending Difficulty

Child cannot write numerals to represent dictated numbers; for example, writes 345 for "three tens, four hundreds, five units."

2.2.5 Problems in Auditory Sequential Memory

1. Child can not state complete basic addition fact, such as "three plus five equals eight." Instead, he forgets—"three plus . . . what did you want me to add?"
2. Child does poorly on repeating a series of digits.
3. Child has difficulty with syntax (order) embedded in simple equations, such as

    $$15 - 7 = \square \, , \, 7 + \square = 15.$$

A major characteristic of psychomotor factors that affect mathematics learning is difficulty in integrating information derived from one sense modality with that gained from others. This is called *sensory integration*. Prominent among this type of difficulty are perceptual and visuomotor disturbances. Birch and Belmont (1965) stated that "many children with specific learning disabilities show themselves to be significantly defective in their ability to integrate information derived from the auditory, visual, tactile, and kinesthetic sense modalities with one another" (p. 295). Since so many mathematical ideas are abstracted from the child's perceived actions upon objects in the real world, the ability for sensory integration is essential to mathematics learning.

2.3.0 Rules of General Language and Mathematics

Rules of general language and examples of how they relate to mathematics include the following:

2.3.1 Phonological Rules

Interpretation and expression of combinations of sounds.

Examples

If a child is unable to discriminate differences or similarities in sounds he is often not able to notice certain quantitative relationships. These include singular-plural relations that are not apparent because of deficit in discriminating the final s of words. Topological notions will be affected, as in the difference between "in" and "on."

(a) "Circle the sets with 3."

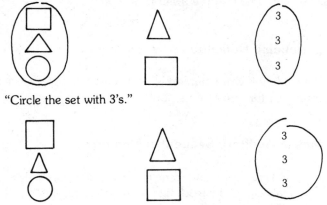

"Circle the set with 3's."

(b) "Circle the correct picture: Stand *on* the box."

2.3.2 Morphological Rules

Govern word structure and inflections; combine the smallest meaningful grammatical units (morphemes). Included here are inflectional endings that form plurals as well as comparatives, superlatives, and differences in tense.

Examples

    (a) a cookie   some cookies
    (b) tall   taller   tallest
    (c) I have . . .   I had . . .

2.3.3 Syntactical Rules

Deal with word order in a sentence, such as subject preceding predicate (babies-cry), adjectives preceding nouns (pretty-lady), or adverbs following verbs (yell-loudly). Syntax, involving order, changes the meaning of sentences that express various mathematical relationships.

Examples

(a) You can put the ball inside the hoop. (enclosure)
(b) You cannot put the hoop inside the ball.
(c) The tallest girl was first in line. (order)
(d) The first girl was tallest in line.
(e) $3 - 2 = 1$ $2 - 3 = -1$
(f) $15 - 3 = 5$ $3 \div 15 = 3/15 = 1/5$

2.3.4 Semantical Rules

Relate the underlying meaning of words.

Examples

(a) If the meaning of "four" is understood, then the child will also know that

$$2 + 2, 5 - 1, 2 \times 2, 8 \div 2$$

and so on is also four.

(b) If multiplication is presented in two ways—as a union of disjoint equivalent sets and as a cross-product of two sets—child will abstract the multiplication idea.

$5 + 5 + 5 = 3 \times 5 = 15$ "Three 5's are 15."
"Two skirts and three blouses form six different outfits."

If a child has an auditory reception impairment, all levels of language function will be affected. Milestones in language development that affect mathematics learning will be impeded. These milestones, summarized by Striffler (1976), are here related to learning mathematics:

• *End of first year:* Child comprehends many simple, concrete, frequently repeated words.
• *End of second year:* Child comprehends some topological notions, such as "Go into the house."
• *End of third year:* Child has good spatial-temporal orientation and comprehension of topological notions expressed as prepositions: "under," "on," "inside."

- *End of fourth year:* Child understands some concepts and responds to directions involving number: "Give me two blocks," "How old are you?" (Child holds up four fingers.)
- *End of fifth year*: Child understands number properties less than 7, identifies at least four colors, and can follow three-part directions: "Pick up the doll, put it in the truck, and bring me the ball."
- *End of sixth year:* Child understands number concepts to 10—"Give me nine marbles"; shows comprehension of direction: "right–left." Demonstrates temporal understanding: "morning–afternoon–night"; understands "more–less," "pair," "many–few."
- *End of eighth year:* Bulk of language learning is complete.

Wiig and Semel (1976) used the Structure of Intellect Model (Guilford, 1967) to characterize language deficits resulting from auditory disabilities. Factors that influence mathematics learning include inability to "comprehend linguistic concepts requiring logical processing such as comparative, temporal, spatial" (p. 30). They also summarized research on language processing disabilities and deficits in mathematics (Wiig & Semel, 1976).

Summary Thought Questions and Activities

I. You should now be aware of psychomotor factors that may affect learning mathematics.
    A. List any of these factors that affected your own mathematics learning.
    B. Write a paragraph on how you have coped with a psychomotor impairment. Include its effect on your own learning and strategies you used to circumvent such impairment.
II. Translate behaviors described by Compton and Bigge in sections 2.1.0 and 2.2.0 and the behaviors in 2.3.0 into assessment activities. For example, for section 2.1.1, show the following operational signs and ask the child to circle the addition sign:

$$- \times \div +$$

III. Describe how you will employ your knowledge of the various generic factors discussed in this chapter in selecting instructional aids and materials.
IV. List all ideas that are new to you as a result of reading this chapter. Compare your list with the chapter contents. Identify ideas you did not remember but that you consider essential in teaching an exceptional child.

Suggested Readings

Aiken, L. R., Jr. Language factors in learning mathematics. *Review of Educational Research,* 1972, *42,* 359-85.

Carrow, M. A. The development of auditory comprehension of language structure in children. *Journal of Speech and Hearing Disorders,* 1968, *33,* 99–111.

Chomsky, C. *The acquisition of syntax in children from five to ten.* Cambridge, Mass. MIT Press, 1970.

Cohen, R. Arithmetic and learning disabilities. In H. R. Myklebust (Ed.), *Progress in learning disabilities* (Vol. 2), New York: Grune & Stratton, 1971.

Dahmus, M. E. How to teach verbal problems. *School Science and Mathematics,* 1970, *70,* 121–38.

Horowitz, R. Teaching mathematics to students with learning disabilities. *Academic Therapy,* Fall 1970, *6,* 17–35.

Kaliski, L. Arithmetic and the brain-injured child. *The Arithmetic Teacher,* 1962, *9,* 245–51.

Linville, W. J. *The effects of syntax and vocabulary upon the difficulty of verbal arithmetic problems with fourth-grade students.* Unpublished doctoral dissertation, State University of Iowa, 1970.

McNeil, D. *The acquisition of language.* New York: Harper & Row, 1970.

Menyuk, P. *The development and acquisition of language.* Englewood Cliffs, N.J.: Prentice-Hall, 1971.

Rosenthal, D. D., & Resnick, L. B. Children's solution processes in arithmetic word problems. *Journal of Educational Psychology,* 1974, *66,* 817–25.

Semel, E. M., & Wiig, E. H. Comprehension of syntactic structures and critical verbal elements by children with learning disabilities. *Journal of Learning Disabilities,* 1975, *8,* 53–58.

Vander Linde, L. F. Does the study of quantitative vocabulary improve problem solving? *Elementary School Journal,* 1964, *65,* 143–52.

References

Abercrombie, M.L.J. Eye movements, perception and learning. In V.H. Smith (Ed.), *Visual disorders in cerebral palsy.* London: Spastics Society/Heinemann, 1963.

Birch, H. G., & Belmont, L. Auditory-visual integration in brain-damaged and normal children. *Developmental Medicine and Child Neurology,* 1965, *7,* 295.

Compton, C., & Bigge, J. Academics. In J. Bigge & P. O'Donnell (Eds.), *Teaching individuals with physical and multiple disabilities.* Columbus, Ohio: Charles E. Merrill, 1976.

Guilford, J. P. *The nature of human intelligence.* New York: McGraw-Hill, 1967.

Hebb, D. O. *The organization of behavior.* New York: Wiley, 1949.

Koupernik, C., MacKeith, R., & Williams, J. F. Neurological correlates of motor and perceptual development. In W. M. Cruickshank & D. Hallahan (Eds.), *Perceptual and learning disabilities in children. Vol. 2: Research and theory.* Syracuse: Syracuse University Press, 1975.

Strauss, A. A., Lehtinen, L. E. *Psychopathology and education of the brain-injured child.* New York: Grune and Stratton, 1947.

Striffler, N. Language function: Normal and abnormal development. In R. B. Johnston & P. R. Magrab (Eds.), *Developmental disorders: Assessment, treatment, education.* Baltimore: University Park Press, 1976.

Wiig, E., & Semel E. *Language disabilities in children and adolescents*: Columbus, Ohio: Charles E. Merrill, 1976.

Physical and Sensory Factors That Influence Learning Mathematics

3

Discussed in this chapter are sensory limitation, low vitality conditions, fatigue, and physical impairments that detract from learning mathematics.

Until recently, children with physical impairments were educated in settings apart from the mainstream. Mullins (1971) pointed out that there is little educational justification for such placement:

> Many teachers have reported an I.Q. range of over one hundred points in their classes. It is not unusual to find a gifted child with muscular dystrophy sitting next to a cerebral palsied child with unmeasured intelligence or vice versa. In the same class there may be children with communication problems such as blindness or deafness who may also be physically handicapped. There will often be one so-called "brain-damaged wall-climber." The age range of such a class may extend from pre-puberty through adolescence. (p. 15)

This situation is changing rapidly as implementation of federal law progresses. Reynolds and Birch (1977) succinctly summarized state and federal litigation that has obligated public schools "to provide appropriate education for literally all children, either in the school's own facilities or by arrangement with other agencies."[1] Public Law 94-142, the Education of All Handicapped Children Act of 1975, embodied the following principles: right to education, least restrictive alternative educational setting, individualized programming, and due process.[2] The principle that most affects regular classroom teachers is the idea of "least restrictive environment" or "mainstreaming."

Special education teachers of the physically handicapped have access to various resources, including doctors, therapists, social workers, vocational counselors, and psychologists. Such teachers have been apprised of materials and methods for educating youngsters with physical disabilities. Academic focus has been on "reading, spelling, handwriting, and thinking skills" (Compton & Bigge, 1976, p. 190). It is interesting to note that mathematics is missing from this list.

The position taken here is that mathematics is a language of relationships communicated via symbol systems. Thus, many of the factors that affect acquisition of reading, spelling, handwriting, and thinking skills may be generalized to mathematics. Factors that influence mathematics learning include those that prevent or inhibit development of spatial and temporal relationships. If the physical handicap prevents body movements, inhibits manipulation of concrete representations of number, or deprives the child from compensating the disability in some manner, then the handicap will make it difficult for the child to develop mathematical ideas.

In order for the teacher to assess mathematics needs of physically impaired students, he or she must understand manifestations of the handicap.

> Younger children build up knowledge of objects in space by information from bodily movements transmitted along various sensory channels. If they are restricted in their activities by motor handicaps, particularly early in life, they would become deprived of a wide range of sensorimotor experiences. . . . Some cerebral palsied children are deprived of opportunities to physically explore the environment, and many suffer an abnormal reduction of preschool experiences. They are unable to manipulate various objects and toys, cannot participate in normal play activities, such as . . . pulling and pushing equipment etc., and seem unable to develop the appropriate . . . concepts of mathematics. (Haskell, 1973, p. 25)

Phillips and White (1964) found that young children with motor handicaps were significantly lower in acquisition of arithmetic skills than were physically handicapped children without congenital brain injury. The time of onset of the physical disability made a difference in acquisition of mathematical performance.

3.1.0 Physical Impairments

Children with physical disabilities[3] may not have had the experiences most children encounter to build a foundation for mathematics (Connor, 1967). These include preparation for applied mathematics such as handling money, buying candy at the store, receiving change, measuring material for a dress, measuring wood for a birdhouse, following a recipe. Connor suggested that the teacher should not assume a knowledge of the value of coins, knowledge of differences in sizes and shapes, or ability to understand mathematical relationships such as big–little, long–short; these situations have no meaning for a child who has not handled objects.

Sattler (1974) summarized research that focused on physical disease or disability and level of intelligence:

> Some generalizations are possible concerning the effects of physical disease or physical disability on the I.Q. The age at which the disease is contracted appears to be directly related to intelligence level. Before ages 3, 4, 5, or 6 years . . . , many conditions result in lowered intelligence. Cardiac conditions after 3 years of age, diabetes mellitus after 5 years of age, idiopathic epilepsy [without known or specific cause] after 5 years of age, and rheumatic fever after 6 years of age seem to have little effect on the I.Q. Conditions which do not appear to affect intelligence include perinatal stress [pertaining to before, during, or after birth] and poliomyelitis. Some effect on intelligence although not very severe, occurs in blindness, deafness, neonatal hyperbilirubinemia [jaundice], and perinatal

anoxia. Severe adverse effects on intelligence occur in cerebral palsy [neuromus-cular impairment], symptomatic epilepsy [a symptom of organic brain damage], very low birth weight, and muscular dystrophy [resulting from a biochemical malfunction].   (p. 345)

It appears that physical factors which most affect mathematics learning are confounded by accompanying intellectual factors. Haskell (1973) sum-marized research on factors affecting arithmetic attainment in cerebral palsied children. These studies offer evidence that inefficiency in searching and in other psychomotor skills inhibits learning mathematics.

3.2.0 Low-Vitality and Fatigue Conditions

Children with low-vitality conditions[4] may be exposed to a mathematics program presented at a slower than usual rate of instruction. This adjustment depends upon how much instruction the child can tolerate. The result is introducing the child to only a portion of the total mathematics curriculum. Thus, selection of appropriate and relevant topics and skills is essential.

Severe weakness may place limitations on amount and degree of movement in space. Chronic medical problems might well result in general weakness. A long-standing medical problem might necessitate severe curtail-ment of activities because the child becomes exhausted in a very short time (Connor, 1967). On the other hand, some children with a chronic medical problem—for example, a congenital heart defect—who reach school age are handicapped little, if at all, by the disability. In general, children with low-vitality conditions due to a chronic medical problem fall within all of the intellectual ranges. Thus, the effect of a low-vitality condition will be determined by concomitant cognitive, sensory, social and emotional factors. The age of onset and the severity of the condition also will be involved in the effect on learning mathematics.

Fatigability is a concomitant factor of both low-vitality conditions and some physical impairments. The child who has difficulty sustaining energy is prevented from attending to academics.

3.3.0 Sensory Limitation

Sensory awareness involves a nonverbal stimulus (light or sound) activating a sense organ (eye or ear) and being detected by the receiver. Striffler (1976) discussed effects of sensory limitation on perception as follows:

Perception is . . . the process of attaching meaning to sensation and involves detection, discrimination, identification, and association of incoming stimuli. . . . Receptively, the infant can and does discriminate between frequencies, inten-sities, and durations at birth. . . . An impairment in sensory awareness . . . will result in a decrease in the quantity or quality of stimulus received. When auditory sensations are lacking or reduced, the child experiences difficulty in relating sounds to experiences. His perceptions, symbols, and concepts are altered. (pp. 78–79)

Sensory impairment of a visual nature affects skills in object identification; discrimination; sorting; classification; quantitative comparisons (size of objects, weight, length, surface measure); orientation and mobility; handwriting; reading temperature; identifying time on a clock face; and using a calendar. Reduced vision limits the child's experiences with her environment, development of spatial orientation, and learning facilitated by imitating others. Unless an alternate symbol system is used, the symbols of mathematics are lost to a visually impaired child. (Alternative systems such as the Braille alphabet will be presented by Chapter 11.)

An auditory impairment results in difficulty in acquiring expressive speech and language. Language is a vehicle for communication of experiences and ideas. Symbols and rules for combining these symbols must be learned in order to use a particular language. The symbols may be words used in discourse, or they may be mathematical.

Summary Thought Questions and Activities

I. You should now be aware of physical factors that may affect mathematics learning.
    A. List any of these factors that affected your own mathematics learning.
    B. Write an essay on how you now feel about a physical impairment of your own and how you have dealt with it. Include its effect on your own learning. Reread this essay in one week and then either share it with a trusted friend, discuss it, or save it to reread at a later time.
    C. List how a physical factor has inhibited your success in learning particular mathematical ideas.
II. List possible effects of low-vitality conditions and fatigue on learning mathematics, and describe these relationships.
III. Write reasons for mathematics difficulties in the hearing-impaired.

Suggested Readings

Aiello, B. *Making it work: Practical ideas for integrating exceptional children into regular classes.* Reston, Va.: The Council for Exceptional Children, 1976.

Bigge, J. L., & O'Donnell, P. A. *Teaching individuals with physical and multiple disabilities.* Columbus, Ohio: Charles E. Merrill, 1976.

Billings, H. K. An exploratory study of the attitudes of non-crippled children in three elementary schools. *Journal of Experimental Education,* 1963, *31,* 381–387.

Buchanan, R., & Mullins, J. B. Integration of a spina bifida child in a kindergarten for normal children. *Young Children,* September 1968, 339–344.

Euphemia, R. Mathematics for the special child. *Academic Therapy*, Fall 1970, 6, 13–15.

Sax, G., & Ottina, J. R. The arithmetic achievement of pupils differing in school experience. *California Journal of Educational Research*, 1958, 9, 15–19.

References

Abeson, A., Bolick, N., & Hass, J. *A primer on due process*. Reston, Va.: The Council for Exceptional Children, 1975.

Compton, C., & Bigge, J. Academics. In J. Bigge with P. O'Donnell, *Teaching individuals with physical and multiple disabilities*. Columbus, Ohio: Charles E. Merrill, 1976.

Connor, F. P. The education of crippled children. In W. M. Cruickshank & G. O. Johnson (Eds.), *Education of exceptional children and youth*. Englewood Cliffs, N.J.: Prentice-Hall, 1967.

Haskell, S. H. *Arithmetical disabilities in cerebral palsied children*. Springfield, Ill.: Charles C. Thomas, 1973.

Mullins, J. B. Integrated classrooms. *Journal of Rehabilitation*, 1971, 37, 14–16.

Phillips, C. J., & White, R. R. The prediction of educational progress among cerebral palsied children. *Developmental Medicine and Child Neurology*, 1964, 6, 167–174.

Reynolds, M. C., & Birch, J. W. *Teaching exceptional children in all America's schools: A first course for teachers and principals*. Reston, Va.: The Council for Exceptional Children, 1977.

Sattler, J. M. *Assessment of children's intelligence*. Philadelphia: W. B. Saunders Company, 1974.

Stevens, G. D. *Taxonomy in special education for children with body disorders*. Pittsburgh: Department of Special Education and Rehabilitation, University of Pittsburgh, 1962.

Striffler, N. Language function: Normal and abnormal development. In R. B. Johnston & P.R. Magrab (Eds.), *Developmental disorders: Assessment, treatment, education*. Baltimore: University Park Press, 1976.

Endnotes

1. The Rehabilitation Act of 1973 provided for funding educational agencies that adhere to the concept of educating *all* children in the most normal setting feasible. This is in line with the notion of *due process*, which insures that each child receive an appropriate education in the least restrictive setting.

2. Effectively complying with due process requirements requires designing educational programs to meet the needs of each child. Determination of both the child's needs and the appropriate program is the purpose of due process. The process, however, is also designed to insure that handicapped children receive their education in the least restrictive setting. Such a concept assumes that there are a variety of alternative settings in which the child can be placed. These settings range from the usual classroom with nonhandicapped peers to more restrictive settings such as special classes on a full- or part-time basis, special schools, or residential institutions (the most restrictive).

Due process frequently is referred to as *mainstreaming*. From a legal view, any removal (as a matter of public policy) of an individual from a normal situation into one that is restrictive is a limitation of that individual's liberty. Thus, placement in a special class as opposed to a regular class deprives the individual of some liberty, and is restrictive (Abeson, Bolick, & Hass, 1975).

3. Stevens (1962) distinguished the terms *impairment*, *disability*, and *handicap*. *Impairment* is the physical defect itself (for example, absence of fingers, diabetes, cardiac disease, or a port-wine nevus—a purple birthmark on the face). *Disability* is a matter of function—a limitation of behavior directly dependent upon the impairment (absence of fingers results in a disability or lack of digital dexterity; cardiac disease inhibits energy levels). A *handicap* is measured by the extent to which an impairment, a disability, or both interfere with learning.

4. Connor (1967) described handicapping conditions and their symptoms involving low vitality. For example, *rheumatic fever:* pallor; loss of weight; nose bleed; poor appetite; low and persistent fever; twitching of face, arms, and/or legs, and pains in the extremities, abdomen, and joints. *Tuberculosis:* general feeling of fatigue on rising in the morning, loss of weight, coughing, indigestion, blood spitting, chest pains, and husky voice. *Diabetes:* excessive thirst; energy is decreased as sugar is withheld; although child eats inordinate amounts of food he may continue to lose weight and strength.

Social and Emotional Factors
That Influence Learning Mathematics

4 Problems of learning mathematics usually relate to either a general learning problem or an emotional difficulty. In our culture, mathematics is considered to involve superior reasoning ability and is thought to be a powerful tool. Successful performance in mathematics carries with it positive connotations. Being "good in math" is "being bright," and being bright in mathematics is associated with control, mastery, quick understanding, leadership. Unsuccessful mathematics achievement implies the opposite of these positive connotations. This value system is a *cultural problem* that has a subtle harmful effect on a number of children and adults.

Parents often give their child the impression that learning mathematics is difficult and that they would like the child to do it. Some may wish that the success in mathematics which eluded *them* be realized in their child. Parental pressure may bear upon general learning, but it is particularly evident for mathematics learning. This is a *parental problem*, and it causes a great deal of "math anxiety" for both parents and child.

Psychoanalytic theory suggests that there is a link between solving mathematics problems and a child's solving his or her own personal problems. Children who don't want to go inside themselves often do not want to analyze problems in mathematics. This is the *child's problem.*

A mathematics teacher must be aware of these external and internal pressures. If a learning problem stems from an underlying anxiety, then a more general kind of help is needed—perhaps psychotherapy. In such instances, the anxiety may be manifested as follows: A child wants to remain a baby because he is afraid to be a man. For the child, mathematics or learning mathematics is symbolic of growing up. Learning mathematics is avoided because of this fear. When *primary* anxiety of this nature is present, the child probably needs help beyond that which a teacher can provide.

On the other hand, a teacher can help a child who displays *secondary* anxiety, which is symptomatic in nature and which may be the result of learning problems. Symptoms may take the forms of hyperactivity, withdrawal, aggression, physical maladies, bed wetting, and so on. This level of anxiety may be exemplified as follows: A child has a perceptual, spatial, temporal, or motor problem, or a combination of these. He or she wants to learn, but cannot produce. The child tries, but the results are inadequate. The child becomes anxious. In this case, the teacher should try to determine what is going wrong at the learning level. Is the work too difficult? Are there gaps in the mathematics curriculum? Does the child need more work at the concrete level? Does a perceptual disability need to be circumvented? Once an instructional problem is diagnosed and corrected, the teacher may be instrumental in decreasing the child's anxiety by providing success situations.

The teacher must keep three factors in mind: the child's rate of intellectual development, the possibility of psychomotor problems, and the distinction between primary anxiety and symptomatic anxiety. All three factors are important when considering emotional and social influences on learning. For example, a bright child who is failing mathematics may have either a psychomotor disability or an emotional disorder, or both, and these may result in social problems. Emotional problems may result from lack of success caused by an undetected visual-perception problem. A slow-learning child may also become frustrated by a perceptual disability. A severely emotionally disturbed child may have mild psychomotor problems that compound the emotional situation. The "whole" child and his interactions with others must be kept in mind.

A distinction is made here between social and emotional factors that influence learning mathematics. Social factors include the following:

- rules of conduct, moral codes, values, customs
- modeling others' behavior, being aware of cues in the environment, relating to and interacting with other people
- using diplomacy, understanding another's point of view, empathizing, enjoying company of others, including in one's decisions the desires and intentions of others
- accepting needed help, forming a balance between autonomy and dependency[1]

Emotional factors include:

- feeling afraid, anxious, frustrated, joyous, angry, surprised
- becoming overly upset, moody, sad, happy

(These influences serve as the basis for mathematics instruction in Chapter 12.)

Social Factors

Social disturbances are often apparent in children who are termed handicapped or talented. Children with special learning needs often display aspects of behaviors listed in the following categories (Myklebust, 1971):

Cooperation. Group participation requires the ability to follow directions without unduly disrupting the activities of others . . . the child may be unable to inhibit his reactions; he may speak out randomly, not wait his turn, or engage in other inappropriate acts. His disruptiveness may be episodic. He may be aware that his behavior is unsuitable, but he is incapable of altering it.

Attention. Inattention is of two major types—distraction from within and distraction from without. The child who cannot control inner distractions is described as disinhibited, whereas the one who overreacts to the surrounding environment is designed as distractible. (Examples of inattention are: never attentive, very distractible; rarely listens, attention frequently wanders.)

Organization. The child lacks facility in planning, cannot organize tasks sequentially, always needs suggestions as to the next step, is careless, inexact.

New Situations. The child shows low tolerance for any change. Some overreact only to particularly stressful situations involving surprise, complex social demands, or fatigue. Related are: excitability, lack of self-control.

Social Acceptance. The child lacks ability to relate well, especially to other children. They are often viewed as unfriendly, disobedient, naughty, to be avoided.

Responsibility. The child is deficient in ability to assume responsibility, lacks general initiative and self-sufficiency.

Completion of Assignments. The child often cannot finish assigned work, even with supervision.

Tactfulness. The child is often rude, often disregarding the feelings of others. This relates to a lack of perception in discerning the wishes of others. This deficiency in social perception precludes learning the significance of cues, especially of certain nonverbal aspects of life. Thus, the child is not aware of the meaning of the actions of others or of himself.   (pp. 9–11)

Another excellent source of factors regarding socialization that influence learning is the American Association on Mental Deficiency (AAMD) Adaptive Behavior Scale (Nhira, Foster, Shellhaas, & Leland, 1974). Socialization involves processes which lead a child into group experiences. Some of these experiences include the following (Wood, 1975): taking turns, sharing, participation in what others suggest, recognizing character traits of others, developing preferences for friends, supporting others, and eventually participating as an invited member of a group. A few examples of AAMD Adaptive Behavior Scale items that relate to learning mathematics are listed in Table 4.1. Statements within a group are numbered in order of difficulty: 4, 3, 2, 1, 0. The directions state to circle the numeral by the statement "which best describes the *most difficult* task the person can usually manage." For some items you are to "check all statements which apply and sum number checked." Other categories include *Responsibility, Socialization, Anti-Social Behavior, Withdrawal,* and *Use of Medications,*[2] among others.

Emotional Factors

Adults can help alleviate fears and anxieties that have an effect upon learning by providing information, reassurance, and comfort. Krutetskii (1976) commented on the role of emotion in learning mathematics:

> The emotions a person feels are an important factor in the development of abilities in any activity, including mathematics. A joy in creation, a feeling of satisfaction from intense mental work, and an emotional enjoyment of this process heighten a person's mental tone, mobilize his powers, and force him to overcome difficulties. An indifferent person cannot be a creator.   (p. 347)

Emotional problems may cause and also result from mathematics difficulty (Reisman, 1978):

Failure in arithmetic can be caused by emotional problems but it also can cause emotional problems. Some children already have developed a fear of arithmetic before they come to school. They may have heard a parent talk about his difficulty or failure in arithmetic and subconsciously may identify with this parent. . . . Sometimes, a parent's occupation is mathematically oriented. Such children may have difficulty in arithmetic either because they would rather fail than compete with their parents, or because they try so hard to succeed that they block their learning.   (pp. 22–23)

Anxiety has generally been regarded as a *response* that may be expressed at various levels of intensity and may involve physiological and motor behaviors such as excessive fidgeting, squirming, blushing, sweaty palms. Investigators have done extensive research regarding the nature of anxiety and its relationship to need achievement, school achievement, test situations, fear, intelligence, and other variables.

Sarason (1958) studied test anxiety and devised two scales to measure anxiety. They are the Test Anxiety Scale for Children and the General Anxiety

Table 4.1. *Selected Categories of the AAMD Adaptive Behavior Scale*

Item	*Category*	
18	Sense of Direction	
	Goes a few blocks from school ground without getting lost	3
	Goes around school ground without getting lost	2
	Goes around school room alone	1
	Gets lost whenever he leaves his own room	0
39	*Social Language Development*	
	Can be reasoned with	——
	Obviously responds when talked to	——
	Talks sensibly	——
	Repeats a story with little difficulty	——
55	*Perseverance-Attention*	
	Will pay attention to purposeful activities for more than 15 minutes	4
	Will pay attention to purposeful activities for at least 15 minutes	3
	Will pay attention to purposeful activities for at least 10 minutes	2
	. . . for at least 5 minutes	1
	Will not pay attention. . . for as long as 5 minutes	0
56	*Persistence*	
	Becomes easily discouraged	——
	Fails to carry out tasks	——
	Jumps from one activity to another	——
	Needs constant encouragement to complete task	——

Scale for Children. The rationale underlying these two scales is that "anxiety is a conscious experience; it can be communicated to another person. Thus, one may determine its occurrence by direct questioning" (Sarason, Lighthall, Davidson, Waite, & Ruebush, 1960, p. 35). Sarason regards anxiety as a stimulus rather than as a response:

> where anxiety is felt by an individual in a learning situation, there are two sets of responses—the task responses which are evoked by the learning situation and which if relevant and correct, will lead to the solution of the task. Secondly, however, the felt anxiety also acts as a stimulus and this evokes its own set of responses (inadequacy, helplessness, somatic reaction) which will interfere with the ongoing and relevant task responses.   (Sarason et al., 1960, p. 39)

Crandall (1967) noted that (a) anxious children tend to perform more poorly on achievement and intelligence tests; (b) anxiety increases with age; (c) girls' anxiety scores are consistently higher than boys (see also Tobias, 1978); and (d) the characteristics of the learning or testing situation must be taken into account.

Biggs (1962) emphasized the importance of identifying the child's level of performance and then giving him tasks that relate to this level in order to avoid frustration that underlies behavior disorder. He stated:

> There is, it seems, usually a very good reason why an individual becomes anxious in the sort of situation in which he is required to do something at a certain level and he is not sure whether or not he will be able to do it—he questions his "cognitive integrity" and sees the situation as being threatening to his own self-esteem.   (p. 120)

Kauffman (1977) discussed four clusters of social and emotional factors including: hyperactivity, distractibility, impulsivity; aggressiveness; withdrawal, immaturity, inadequacy; and deficiencies in moral development.

4.1.0 Hyperactivity, Distractibility, Impulsivity

These three factors, although involving different behaviors, are interrelated and occur frequently in those students with emotional disturbance, learning disabilities, and mental retardation. They have been delineated by Kauffman (1977, p. 146) as follows:

> *Hyperactivity* or *hyperkinesis* involves excessive motor activity of an inappropriate nature.
> *Distractibility* is the inability to selectively attend to the appropriate or relevant stimuli in a given situation, or overselectivity of attention to irrelevant stimuli.
> *Impulsivity* is disinhibition or a tendency to respond to stimuli quickly and without considering alternatives.

4.2.0 Aggressiveness

Group interaction in a classroom requires cooperation among members of the social environment. Rules of conduct that facilitate learning involve following directions, taking turns and, in general, not being a disruptive force. The child who is unable to inhibit verbal or motor activity will have difficulty in engaging in suitable behavior. Bandura (1973) listed considerations for identifying aggressive behavior, including evaluating (a) the behavior itself, (b) apparent intentions of the aggressor, and (c) type of reactions on the part of the target of the behavior.

4.3.0 Withdrawal, Immaturity, Inadequacy

The Developmental Therapy Model (Wood, 1975) lists the three D's of behavior disorder. These include:

* *Depth*—How unusual or severe is the child's problem?
* *Duration*—How long has it been going on? Since birth? Only in mathematics? Only on the playground?
* *Development*—Does the problem interfere with the child's development? What is the impact on learning?

The child who has a severe problem would be withdrawn to the point of being an *isolate, nonentity, wallflower,* or *hermit in residence.* This condition would be debilitating to the child's development.

4.4.0 Deficiencies in Moral Development

Kohlberg (1969) presented a stage theory of moral development based upon the developmental theory of Piaget. There are three main stages:

* *Preconventional morality* where children's behavior is controlled by external forces that involves obtaining material rewards and fear of punishment. Later during this stage children become less concerned with physical rewards and punishments and become more concerned with pleasing significant others (parents, teachers, authority figures). Concepts of reciprocity and sharing emerge and children begin to consider reactions of others to their own actions.
* *Conventional morality* involves self-control and conscious need for acceptance and social approval. Later in this stage morality takes on a legal aspect—a law and order morality.
* *Postconventional morality* involves making independent moral judgments.

Biehler (1974, p. 395) presented a concise comparison of Kohlberg's stages of moral development and Piagetian theory:

> Just as Piaget argues that the child advances through the stages of intellectual development in sequence, Kohlberg maintains that children must pass through lower stages of moral development in order to become capable of postconventional thinking. . . . The Kohlberg stages are not clearly related to specific age levels, particularly since many individuals never progress beyond the conventional level. However, the switch from preconventional to conventional thought typically takes place around the end of the elementary grades.

Summary Thought Questions and Activities

I.   You should now be aware of social and emotional factors that may affect mathematics learning.
    A.   Which of these factors affected your own mathematics learning?
    B.   Which of the social and/or emotional factors discussed in this chapter enhanced your own educational experiences in mathematics?
    C.   Which inhibited your success in mathematical reasoning?
    D.   If possible, verify your perceptions with family, former teachers, friends.
II. Describe how you can employ these generic factors to avoid causing behavior problems in your students.
III. List all ideas that are new to you as a result of reading this chapter. Compare your list with the chapter contents. Identify ideas you did not remember but that you consider essential in teaching an exceptional child.
IV. List additional social and/or emotional factors that, in your experience, affect learning mathematics.

Suggested Readings

Arieti, S. The structural and psychodynamic role of cognition in the human psyche. In S. Arieti (Ed.), *The world biennial of psychiatry and psychotherapy* (Vol. 1). New York: Basic Books, 1970.

Arieti, S. *The interpretation of schizophrenia.* New York: Basic Books, 1974.

Arieti, S., & Bemporad, J. *Severe and mild depression: The psychotherapeutic approach.* New York: Basic Books, 1978.

Bender, L. Genesis of hostility in children. *American Journal of Psychiatry,* 1948, *105,* 241–245.

Bettelheim, B. Listening to children. In P. A. Gallagher & L. L. Edwards (Eds.), *Educating the emotionally disturbed: Theory to practice.* Lawrence, Kansas: University of Kansas Press, 1970.

Compton, M.F. *A source book for the middle school.* Athens, Ga.: Education Associates, 1978.

Cruickshank, W. M., Paul, J. L., & Junkala, J. B. *Misfits in the public schools.* Syracuse: Syracuse University Press, 1969.

Forman, G. E., & Kruschner, D. S. *The child's construction of knowledge: Piaget for teaching children.* Monterey: Brooks/Cole, 1977.

Galvin, J. P., & Annesly, F. R. Reading and arithmetic correlates of conduct problem and withdrawn children. *The Journal of Special Education,* 1971, *5,* 213–219.

Hallahan, D. P. Distractibility in the learning disabled child. In W. M. Cruickshank & D. P. Hallahan (Eds.), *Perceptual and learning disabilities in children volume 2: Research and theory.* Syracuse: Syracuse University Press, 1975.

Haring, N. G. Social and emotional behavior disorders. In N. G. Haring (Ed.), *Behavior of exceptional children: An introduction to special education.* Columbus, Ohio: Charles E. Merrill, 1974.

Hill, K. T., & Eaton, W. O. The interaction of test anxiety and success: Failure experiences in determining children's arithmetic performance. *Developmental Psychology,* May 1977, *13,* 205–211.

Hoch, P. H., & Zubin, J. *Psychopathology of childhood.* New York: Grune & Stratton, 1955.

Osborn, D.K., & Osborn, J.B. *Discipline and classroom management.* Athens, Ga: Education Associates, 1977.

Palermo, D. S., Casteneda, A., & McCandless, B. R. Anxiety and complex learning. In W. J. Meyer (Ed.), *Readings in the psychology of childhood and adolescence.* Waltham, Mass.: Blaisdell Publishing Co., 1967.

Pasanella, A. L., & Volkmor, C. B. *Coming back . . . or never leaving.* Columbus, Ohio: Charles E. Merrill, 1977.

Reynolds, C. R., & Richmond, B. O. What I think and feel: A revised measure of children's manifest anxiety. *Journal of Abnormal Child Psychology,* 1978, *2,* 271–280.

Reusch, J. Social disability: A review. In S. Arieti (Ed.), *The world biennial of psychiatry and psychotherapy* (Vol. 2). New York: Basic Books, 1973.

Sarason, I. G. Empirical findings and theoretical problems in the use of anxiety scales. *Psychological Bulletin,* 1960, *57,* 403–415.

Sarason, S. B., Davidson, D. S., & Blatt, B. The preparation of teachers: An unstudied problem in education. New York: Wiley, 1962.

Shantz, C. The development of social cognition. In E. M. Hetherington (Ed.), *Review of child development research* (Vol. 5). Chicago: University of Chicago Press, 1976.

Szasz, T.S. The myth of mental illness. *American Psychologist,* 1960, *15,* 113–118.

Webb, T. E. Stereoscopic size discrimination in disturbed and nondisturbed children. *Perceptual and Motor Skills,* 1973, *36,* 323–328.

References

Association for Childhood Education International. *Children and drugs.* Washington, D.C.: ACEI, 1972.

Bandura, A. *Aggression: A social learning analysis.* Englewood Cliffs, N.J.: Prentice-Hall, 1973.

Biehler, R.F. *Psychology applied to teaching, (2nd ed.).* Boston: Houghton Mifflin Co., 1974.

Biggs, J. B. Anxiety, motivation, and primary school mathematics. *The National Foundation for Educational Research in England and Wales, Occasional Publication No. 7.* Published by the Information Service of the National Foundation for Educational Research in England and Wales, 1962.

Crandall, V. C. Achievement behavior in young children. In W. C. Hartup & N. L. Smothergill (Eds.), *The young child: Review of research*. Washington, D.C.: National Association for the Education of Young Children, 1967.

Erikson, E. H. *Childhood and society*. New York: W. W. Norton & Co., Inc., 1963.

Kauffman, J. M. *Characteristics of children's behavior disorders*. Columbus, Ohio: Charles E. Merrill, 1977.

Kohlberg, L. Stage and sequence: The cognitive-development approach to socialization. In D. Goslin (Ed.), *Handbook of Socialization Theory and Research*. Chicago: Rand McNally, 1969.

Krutetskii, V. A. *The psychology of mathematical abilities in schoolchildren* (J. Kilpatrick & I. Wirszup, eds., and J. Teller, trans.). Chicago: University of Chicago Press, 1976.

Myklebust, H. R. *The pupil rating scale: Screening for learning disabilities*. New York: Grune & Stratton, 1971.

Nhira, K., Foster, R., Shellhaas, M., & Leland, H. *Adaptive behavior scales: Manual* (rev. ed.). Washington, D.C.: American Association on Mental Deficiency, 1974.

Reisman, F. K. *A guide to the diagnostic teaching of arithmetic* (2nd ed.). Columbus, Ohio: Charles E. Merrill, 1978.

Sarason, S. B. A test anxiety scale for children. *Child Development*, 1958, *29*, 105-113.

Sarason, S. B., Davidson, D. S., Lighthall, F. F., Waite, R. R., & Ruebush, B. K. *Anxiety in elementary school children: A report of research*. New York: Wiley, 1960.

Tobias, S. *Overcoming math anxiety*. New York: Norton, 1978.

Wood, M. M. *Developmental therapy*. Baltimore: University Park Press, 1975.

Endnotes

1. See Erikson (1963) for a full discussion of balance in dealing with life crises and of Freud's states of personality development.
2. See the Association for Childhood Education International (1972) for an overview on the use of stimulant drugs in treating behaviorally disturbed children and for a list of drugs that elementary teachers are likely to encounter.

Instructional Strategies

SECTION II

This section of the book deals with curriculum goals and methods of teaching these goals to children with special needs that affect their learning mathematics. These needs, discussed in Section I, deal with the cognitive, psychomotor, physical, sensory, social and emotional aspects of learning mathematics. "Instructional Strategy" (IS) refers to modifications in both curriculum and methodology. Curriculum is an educational term meaning what is taught; a method of instruction describes how something is taught.

Curriculum Considerations

Curriculum concerns involve qualitative decisions about the mathematics content that is taught. The degree of severity of the handicapping condition(s) will have a great effect upon what curriculum is appropriate for what child. For example, if cognitive factors are considered, the mathematics curriculum will be selected primarily with regard to the child's rate of cognitive development relative to that of most of his age peers. The greater the retardation in rate of cognitive growth, the less mathematics is learnable for a particular child in terms of abstractness, complexity, and amount. It becomes imperative, therefore, that the child's teachers (and these may include parents, siblings, homebound teacher, residential setting personnel) select the mathematics which is of most use for the child's life needs. For the severely retarded child, the mathematics curriculum may comprise only one goal, one-to-one correspondence. The one-to-one relationship underlies ability to button clothing; sit on a chair; place a piece of food into the mouth and not into eyes, nostrils, or ears; and other basic self-help skills.

For the child whose rate of cognitive growth is accelerated, the goals comprising a mathematics curriculum would be greater in number and much more comprehensive. For each child along the cognitive continuum, the amount of mathematics that may be learned will depend upon the rapidity of the child's learning process. The quality of his thinking, as described by Piagetian developmental learning theory and the literature on creativity, must also be considered.

Very much related to the amount of mathematics that a child may learn is the character of the mathematics curricula. The quality of what is taught must be modified in terms of abstractness of materials, necessary memory load, need for conceptualizing and generalizing, number of ideas presented in a given unit, and complexity of syntactic and semantic aspects of the mathematics material. Instruction that accommodates cognitive factors is discussed in Chapters 5 through 9.

Curricular decisions based primarily upon physical considerations will generally relate more to restrictions resulting from the physical impairment than to mental age. An immobile child may not have a grasp on mathematics relationships that emerge from knowledge of one's movement in space. Therefore, these relationships must somehow be built into this child's curriculum. If necessary spatial relationships cannot be incorporated into the child's repertoire or circumvented, then an alternative selection of mathematics content may be required. If a psychomotor or sensory impairment has prevented acquisition of certain language relationships, such as prepositions and relational adjectives, the mathematics curriculum will also be affected. The curriculum for a physically handicapped or a sensory-impaired child who is older must include systematically presented experiences that are basic to mathematics. Examples of these are: one-to-one correspondence; one-to-many correspondence; cardinality or the "howmuchness" of a set; sequence or ordinality; equivalence of sets or shapes; and topological notions including proximity, separateness, enclosure and order. Chapters 10 and 11 include mathematics instruction appropriate for a child with psychomotor, physical, or sensory impairment.

Mathematics content appropriate for children affected by social and emotional factors is suggested in Chapter 12. High-interest topics are needed as a vehicle for teaching mathematics ideas to unmotivated, withdrawn, socially immature students. Once again, the degree of severity of disturbance will determine the nature of the curriculum adaptation. In the case of a severely disturbed child, for example, it may be appropriate to have no mathematics curriculum until the crisis condition is alleviated.

Method Considerations

Methodology concerns include attention to: quantity of what is taught; pacing that accommodates short attention span and fatigue; rate of presentation; mode of representing the ideas to be taught (for example, concrete, picture, symbols); sequencing of topic; amount of drill and practice; selection of materials appropriate in level of interest and difficulty; development and use of problem-solving skills; and utilization of aids designed to teach the sensory-impaired and physically handicapped.

Methodological strategies are intertwined with curriculum strategies throughout Section II. Both of these categories are included in order to facilitate implementation of curriculum in classroom situations. For an exceptional child in particular, content that is learnable often determines methodology; by the same token, appropriate methods and/or materials sometimes direct content selection.

Section II is structured upon mathematics content for those in the mainstream, with emphasis on kindergarten through grade eight curriculum. This content encompasses the following mathematical ideas:

- *Prenumber relationships*
- *Geometric relationships*
- *Mathematical language*
- *Number*
- *Notation*
- *Operation on sets and on numbers*
- *Applied mathematics*

The underlying question is: As the student processes the mathematics, what is the nature of the student's thinking? Is there an arbitrary association to be made? Does the learner need to grasp relationships among parts? Does she need to apply a relationship across situations? Does he need to conceptualize? Are broad, more abstract relationships or generalizations involved? The teacher must know the psychological nature of what he or she is teaching.

Another consideration for arranging content of Section II in this manner is that the psychological nature of a topic may change as a function of the complexity of a particular task. For example, tasks dealing with a form of seriation increase in complexity from a basic relationship to a higher-level generalization. The experience of the learner is also a factor. The more practice a child has fixing basic addition facts after she has learned them as higher-level generalizations, the sooner they become arbitrary associations at an automatic recall level. Some topics (in particular, basic facts and place value), if taught initially as though they were rote associations instead of generalizations, become of limited utility for the learner in further mathematical endeavors.

Being aware of the psychological nature of our mathematics curriculum is important for all children, but it is imperative when teaching exceptional children.

The instructional strategies derive from research on learning theory and are here applied to mathematics instruction. Some of these are more appropriate to (a) a child operating at a less developed level of cognition, (b) gifted children, (c) a child with social or emotional problems, and (d) those with a physical or sensory impairment. These strategies are not exhaustive, but we hope that they will provide a good beginning for teaching mathematics to a child with special learning needs.

Instead of attempting to apply all instructional strategies to all of the mathematics goals, specific instructional strategies are suggested for selected mathematics content. These selections were based upon the following criteria:

1. appropriateness to the psychological nature of the mathematics content
2. appropriateness to use in regular classrooms for purposes of mainstreaming
3. generalizability to all learners of mathematics as need is indicated
4. generalizability to other components of the developmental mathematics curriculum (see Figure 1.4 in Chapter 1)

Research dealing with teaching mathematics "differently" to different types of learners is unresolved for "normals" and is sparse and unsupported for exceptional children. Therefore, although the strategies in this book stem from learning theory research, implications for teaching are based upon common-sense instruction for the special educational needs of an individual child. The reader should (a) be aware of the nature of mathematics to be taught, (b) use these instructional strategies when diagnostic data on a child suggest, (c) be flexible and try alternative instructional strategies with appropriate mathematics content,[1] (d) discard strategies that do not work for a particular child and try others, and finally (e) not give up but instead create your own instructional modification for your student.

---

1. See Diagnostic Guide, Appendix H

Included in this section are assessment and teaching activities for each major topic. The activities serve two purposes. The assessment activities may be used diagnostically to assess a pupil's mathematical performance and competence in the particular mathematics content. The teaching activities involve instructional strategies. These may also be used for purposes of assessment.

Instructional Strategies Based on Cognitive Needs: Basic Relationships and Arbitrary Associations

5 The ability to construct relationships is a human phenomenon that occurs long before a child enters formal instruction. Arbitrary associations represent relationships as well as more complex mathematical ideas. Children will already know some arbitrary associations before they enter school—for example, the word names *one*, *two*, and *three*, and perhaps their symbols *1*, *2*, and *3*. They may not, however, be able to match these with the numbers they represent. This is obvious when a child counts in either of these ways:

(a) "one, three, five, sixty-million, six"

(b)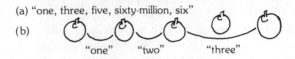
        "one"   "two"   "three"

In example (a), the child has not yet learned the correct verbal chain or succession of number names, similar to learning the letters of the alphabet in accepted sequence. Thus, the ability to sequence precedes the skill of counting. In example (b), a one-to-one relationship is lacking between the number symbol (name and/or digit) and the object counted. In this situation, the one-to-one relationship must be applied to enumerating objects in a set.

Relationships are the foundation for conceptualizing and generalizing. The rate at which a child develops relationships and the kind of relationships a child constructs determine the abstractness and complexity of mathematics the child can learn at various stages of cognitive development. One way of viewing a child's readiness for such basic mathematics ideas as number, shape, size, length, and weight is to assess his ability to sequence and to place sets in one-to-one correspondence. These ideas, as well as the topological concepts of proximity, enclosure, separateness, and order, are fundamental for a developmental mathematics curriculum.

5.0.0 Prenumber Relationships

Sequencing and one-to-one correspondence are basic to competence in mathematics. These concepts will be reviewed and relevant instructional strategies suggested. The child who cannot sequence and who is not aware of one-to-one correspondences will not be able to understand more abstract

51

mathematical ideas. It is important, therefore, to determine whether an older child whose intellectual development is retarded was given the opportunity to learn these prenumber relationships. Since most children develop these at about the time they enter first grade, a child with a mental age of about 6 or 7 should be able to learn them. For the young child with accelerated intellectual development, it is also necessary to insure their awareness of these basics (see Cronbach & Snow, 1977, for a discussion of rates of learning).

5.1.0 Sequencing

Succession signifies one object or event after another without any logical ordering or pattern. *Sequence* designates a following in space, time, or thought within some logical system. A *series* is an arrangement according to a pattern of like relationships with emphasis on each component, rather than one merely following another.

Succession, then, seems to underlie abilities to sequence and to seriate. Succession is merely the act of following in order. Can the child move his feet, one after the other, so that he accomplishes the act of walking? Succession is involved in many of the self-help skills (swallowing before stuffing more food into one's mouth, disrobing before bathing, removing or unzipping one's pants before urinating). Succession also underlies systematic perceiving, in which we "notice" first one object and then another.

Sequence—the process of following in space, time, and thought— implies a simple succession. Tasks involving visual or auditory sequences are considered in the class of verbal learning (see Blake, 1976, for discussion of verbal learning). The following activities assess one's ability to sequence.

Assessment
Activities:

1. Have child follow simple directions as follows:
    a. Put on your socks, then your shoes.
    b. Stand up, jump, sit down.
    c. Clap your hands, stamp your foot, put your hands on your knees.
    d. Connect dots by proceeding from one dot to next to form a figure.

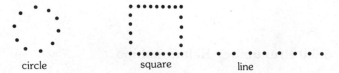

circle   square   line

    e. Count out loud until I say, "Stop."

2. Arrange objects in space
   a. Give the child some buttons and observe if he makes a row, circle, or some other design or if he merely opens and closes his hand amidst the buttons as in kneading bread. This distinguishes between attending/not attending to elements of a set in succession.
   b. Play hopscotch.
   c. Run baseball bases in order.
   d. Copy bead patterns on a string.

3. Sequence in time
   a. Have child imitate your succession of arm movements.
   b. Ask child to describe steps in a game to another child.
   c. Bounce a ball fast, then slowly, then fast. Ask child to bounce ball just as you did.

4. Sequence in thought
   a. Repeat parts of face. Teacher lists mouth, eyes, nose, and so on and asks child to repeat. For example:
      1. eyes, mouth
      2. nose, eyes, forehead
      3. chin, eyes, nose, eyebrows
      4. cheeks, lips, nose, mouth, eyes
   b. Digit span test.[1] Teacher says digits in sequence and child is asked to repeat them. For example:
      1. 2 3
      2. 4 2 6
      3. 7 1 9 6
      4. 5 8 6 4 3
   c. Repeat objects in room, starting with two and continuing with several.
   d. Repeat chain of girls' names, alphabet, streets, children in class, and so on.

Teaching Activities

> ### Instructional Strategy 5.1.1:
> *Present small amounts of the sequence to be learned in an organized format in order to facilitate retrieval.*

Separate the components of simple directions into small bits. For example, modify the directions "Count out loud until I say STOP" to "Count one, two." Then after the child does this, proceed with the direction "Count one, two, three."

> ## Instructional Strategy 5.1.2:
> ### Incorporate redundancy into written presentation.

Have the child write (1, 1, 2, 2, 3, 3) or (1, 2, 3; 1, 2, 3).[2] The repeated sequences may be emphasized by using voice cues if presented orally, or by spacing, underlining, or using color if presented graphically.

You may wish to combine this instructional strategy with the preceding one, employing both reduced amount of material presented and redundancy. For example, instead of directing a child to "Count out loud until I say STOP," give the following commands:

(a) "Count one, two; one, two; one, two."
(b) After the child performs step (a), say, "Count one, two, three; one, two, three; one, two, three."

> ## Instructional Strategy 5.1.3:
> ### Use visual or auditory cues that highlight the position of an item in a sequence.

This approach sometimes facilitates learning of the entire sequence. Examples include writing in a different color or on a different background the digit whose position is either first in a series or in the middle. Mercer and Snell (1977, p. 21) extended this strategy to tasks of auditory sequencing:

> In order to facilitate memory of the entire sequence, a series of words or commands could be broken up by volume or voice variation of one word or command in the primary or middle position. . . . Another variation . . . includes the use of verbal commands for attention nested within the early or middle portions of a list to be recalled: ball, baby, NOW LISTEN, bone, bat, bunny, blue. Although nested attention commands have been used . . . to facilitate recall of the specific item they precede, the commands used here break the sequence into memorable segments.

This strategy may be applied to mathematics instruction. Insert an attention command within an oral sequence: "One, two, LISTEN TO ME, three, four."

The following serve as activities for learning sequences of digits and number names:

(a)   Write a single digit (0-9) on a series of white cards.

0 1 2 3 4 5 6 7 8 9

(b)   Then highlight the digit in either the first or the middle position by substituting another color card or digit in a sequence as follows:

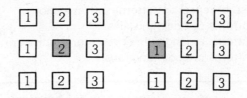

(IS)

Instructional Strategy 5.1.4:
Use separating and underlining as cues.

Present a span of digits in either of the following ways to incorporate separating and/or underlining as aids to memory:

  (a) 1 2 3   4 5 6   7 8 9
  (b) 123456789; 123456789; 123456789;

(IS)

Instructional Strategy 5.1.5:
Control number of dimensions that define a linear sequence.

If the child is expected to sequence by length, then straws, rods, and so forth should differ *only* in length. Do not introduce different colors, textures, or widths into initial sequencing-by-length tasks; that is, omit features irrelevant to the linear sequence.

(IS)

Instructional Strategy 5.1.6:
Emphasize patterns that generate the sequence.

In research on discrimination of relations among sets of stimuli, "with symmetrical arrangements [definite pattern in the stimuli] the retarded and normal subjects performed equally well. With random arrangements, the normal exceeded the retarded" (Blake, 1976, p. 325). Thus, in learning a song, a song that has a sequential pattern (*Row, Row, Row Your Boat*) would be easier to learn than a song with a melody of a nonrepetitive sequence of notes, words, or tones (*The Star Spangled Banner*). The use of pattern as a cueing technique may be used in more complex situations, such as the following:

1.   Use of dominoes to teach number sequences

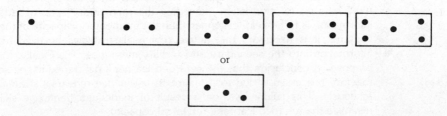

or

2. Use of repeated patterns in translations
3. Patterns of brick and tiles
4. Patterns among the set of natural numbers
5. Patterns in body functions

> ### *Instructional Strategy 5.1.7:*
> *Introduce additional relevant dimensions for generating an alternating pattern.*

Use simple sequential patterns incorporating differences of size, color, and shape. This also serves to develop skill in copying. Note that in this IS relevant dimensions are added, while in IS 5.1.5 irrelevant dimensions were excluded.

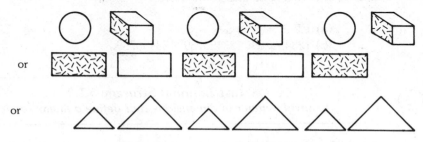

> ### *Instructional Strategy 5.1.8:*
> *Teach use of rehearsal strategies.*

Waugh and Norman (1965, p. 92) described rehearsal strategies as follows:

> rehearsal simply denotes the recall of a verbal item—either immediate or delayed, silent or overt, deliberate or involuntary. The initial perception of a stimulus probably must also qualify as a rehearsal. Obviously a very conspicuous item or one that relates easily to what we already learned can be retained with a minimum of conscious effort. We assume that relatively homogeneous or unfamiliar material must, on the other hand, be deliberately rehearsed if it is to be retained.

Mercer and Snell (1977, pp. 15–16) summarized 14 studies by Ellis (1970) on memory processes of mentally retarded learners:

> Primary and secondary memories are involved when the short-term memory must hold on to information exceeding the momentary span of attention. If information is to be held for longer than the momentary capacity of the primary memory, it is necessary that active rehearsal strategies be used to process information into the secondary and tertiary memories. . . . Finally, Ellis (1970) proposes a conclusion that the retarded learner's deficiency in the secondary memory processes is due to a failure to use active rehearsal strategies spontaneously. This failure may be a result of inadequate language skills which minimize the retarded learner's rehearsal capacity.

Mercer and Snell (1977, pp. 23–24) listed rehearsal strategies that included the following:

1. use of cumulative rehearsal (that is, rote repetition—either aloud or silently—of increasing stimuli including naming the current visual stimulus and then rehearsing three times in sequence all previous items plus the current item)
2. visual imagery strategies (such as use of mental pictures)
3. repetition
4. verbal elaboration
5. systematic scanning and pointing
6. cue selection (for example, use of color or some other dimension of a stimulus)
7. chunking (for example, grouping of material, as in a number series)

Following are examples of rehearsal strategies 1 through 5. Cue selection was discussed in IS 5.1.3, and chunking in IS 5.1.1.

Cumulative rehearsal, visual imagery and repetition. Have child use number word cards and say aloud the number names in the following manner. Present one word card at a time. Child then returns to card one on each group of subsequent trials and proceeds by saying aloud the sequence. He should repeat each grouping three times before adding a new card.

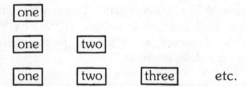

Child reads aloud, "One; one, two; one, two, three" and so on. Then remove the word cards and tell child to repeat this activity by seeing the cards in his mind.

The mentally slower child will need more practice to enable him to retain sequences in time. Periodically teaming children who can help each other (with sequencing or other activities) frees teacher time for other children.

Verbal elaboration. The child listens to the teacher verbalize a sentence about the sequence. For example, for the sequence "stand up, jump, sit down," the teacher performs the acts and says "I stand up, I jump, and now I sit down."

Systematic scanning and pointing. Present the child with a visual sequence of size relationships represented by a picture of different-colored blocks. Instruct the child to point to each of the blocks with her index finger in a left-to-right direction.

Have her repeat this three times. Then give her blocks like those in the picture (such as a medium-size blue block, a small wooden color cube, and a large red block). Ask her to put the blocks in the same order as they are in the picture. You may try this with the picture present and also with the picture removed. You may also represent this sequence with another set of real blocks cemented to an oak tag or a piece of board. This allows for representation of the sequence in two modes—concrete and two-dimensional.

> ### Instructional Strategy 5.1.9:
> *Provide sequencing activities comprised of more complex or subtle nature.*

Many activities incorporating sequencing will have occurred long before a gifted child enters formal schooling. Parents should be encouraged to play games with their children involving pointing in succession and simultaneously naming; for example, parent points to his or her nose, mouth, ear, while naming these features, and then asks child to do the same. Of course, a child can identify a facial feature before he can point and name, but the very young gifted child should do both pointing and naming earlier. This activity is the forerunner of enumerating.

In the preschool and kindergarten, a child is surrounded by sequencing phenomena in her environment. The teacher should serve as a model for engaging in alert, keen observations. Torrance (1970, p. 44) suggests ordering "sequences of events in cartoons, photographs, . . . events in creative dramatics, the daily schedule."

> ### Instructional Strategy 5.1.10:
> *Incorporate incompleteness to activate creative potential.*

Torrance (1970, p. 7), discussing creative and problem-solving activities, encouraged the use of incompleteness in "pictures, stories, objects of instruction, teacher or pupil questions, the behavioral settings of the classroom or playground, or in structured sequences of learning activities." The classroom teacher can develop his or her own creative problem-solving skills at the same time as creative potential is stimulated in the student. The point is that opportunities for using creative potential need to be provided.

Just as certain variables mask giftedness in intelligence (cultural differences, language experience with tasks), circumstance may also mask creative potentialities, primarily "learning by authority." Torrance (1970, p. 2) stated:

> Learning by authority occurs when the learner is told what he should learn and when he accepts something as true because an authority says that it is. . . . Learning by authority appears primarily to involve such abilities as recognition, memory, and logical reasoning—which are, incidentally, the abilities most frequently assessed by traditional intelligence tests and measures of scholastic aptitude. In contrast, learning creatively through creative and problem-solving

activities, in addition to recognition, memory, and logical reasoning, requires such abilities as evaluation (especially the ability to sense problems, inconsistencies, and missing elements), divergent production (e.g. fluency, flexibility, originality, and elaboration), and redefinition.

Torrance (1970, p. 7) presented "strategies for creating and/or using incompleteness to motivate the learning processes and keep them going." Listed are those that have worked particularly well in teaching sequencing, along with an example for each:

1. *Predictions from limited information required.* Play a sequence of patterned beats and ask the student to beat out the next portion of the sequence. Accept patterns other than those expected, as creative persons often come up with original responses.
2. *Tasks structured only enough to give clues and direction.* Have child arrange a comic strip in a new sequence and make up a new story about it.
3. *Encouragement to take the next step beyond what is known.* Have child ask questions about missing parts of a sequence. For example, begin a bead pattern that does not yet make apparent the desired sequence, and play a game in which the student asks questions for guidance. The teacher gives only yes or no answers.

Will they all be the same color? No.
Will they all be circles and squares? Yes.

From an assorted pile of beads of different shape, size, color, and material, encourage child to experiment by adding to the sequence one possibility at a time. For example, the child next adds a small white round bead similar to the first in the sequence:

Is this the next bead? No.
Will some be larger? Yes.

Child then selects a large round white bead.

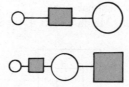

Is this right? Yes.
And this? Yes.
Now do I begin again? No.
Is the pattern done? Yes.

Of course, two children can play this, thus freeing the teacher for other instruction. However, too often creative or gifted children are shortchanged on teacher's time and left too much on their own. Strike a happy balance; avoid instructional strategies for the gifted that involve having them work through their basal text, page by page, independently for the entire year. This often leads to the developing of gaps in their mathematical foundation, and blocks them later on.

> ### Instructional Strategy 5.1.11:
> ### Producing something and using it.

Have the child construct a tessellation out of colored blocks and then describe the various patterns that emerge upon further examination.

5.2.0 One-to-one Correspondence

Matching one entity with another, one-for-one, is basic to understanding mathematics. This relationship may not be apparent to the child who is developing at a retarded cognitive pace. Therefore, experiences that emphasize one-to-one correspondence must be arranged systematically. The pervasiveness of the one-to-one mathematics idea was apparent even when the cave man kept track of the number of his sheep by matching each to a pebble. A theater owner sells one ticket for each seat for a particular performance; only one child sits at a desk in class; one note on a music staff represents one musical sound; and the digits 0–9 each represent a unique number. One-to-one correspondence is basic to arithmetic and geometry—to equal sets and equivalent sets, and to congruence and similarity.

*Assessment
Activities:*

1. Match shapes on simple form boards.

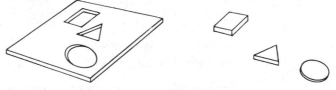

2. Have child bounce a ball as another child claps, jumps, counts, says alphabet, etc.
3. Match blocks to their pictures.

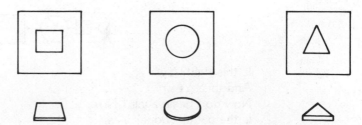

4. Match cowboys to horses, cups to saucers, etc.
5. Put one foot in one pants leg when dressing.
6. Match number name to objects counted in counting sequence.

Teaching Activities

One-to-one correspondence may be exemplified by paired-associates learning tasks. Blake (1976, p. 274) defined paired-associates learning in this way: "the pupil pairs, or learns the relations between particular stimuli and responses. For example, a pupil uses paired associates learning when he learns synonyms, numeral-number correspondences, ..." Conditions that affect verbal learning include characteristics of the material (such as meaningfulness, intralist similarity, serial position) and methods of presentation (for example, organization, prompting, study-test) (Blake, 1974, 1976).

> ### Instructional Strategy 5.2.1:
> *Employ frequency to develop number word–object associations.*

At every opportunity, match up the number words—especially "one" and "two"—with objects. "I have one cookie, I have two cookies, I have three cookies" instead of "I have some cookies." Have the child point or touch each object as he says the sequence of words. Use the word "one" instead of the indefinite article "a"—for example, "I have one car" instead of "I have a car."

> ### Instructional Strategy 5.2.2:
> *Employ familiarity to develop number word–object associations.*

Label sets of objects with different number qualities; for example, sets with "one" color, "one" sex, "one" door, "two" legs, "four" legs, etc. Concentrate on one word at a time. You may also wish to associate the digit with the number word.

two color blocks
2 color blocks
2 colors
2 blocks

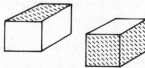

> ### Instructional Strategy 5.2.3:
> *Employ vividness or imagery to develop number word–object associations.*

Tell the child to picture his one closest friend, his two closest friends in his mind.

61

> *Instructional Strategy 5.2.4:*
> *Emphasize patterns that generate the sequences to be matched.*

Match a set of dominoes to a set of playing cards.

5.3.0 Topological Relationships

A child's earliest mathematical discoveries involve concern with spatial relationships that ignore metric properties of Euclidean space. Children first notice topological properties, including proximity, enclosure, separateness, and order. Many of these relationships are expressed as adverbs and prepositions. The child who is unable to express his ideas verbally without difficulty may be able to show you that he understands the difference between *nearness* and *farness*, *within* and *outside* of a boundary, a bicycle tire that is *intact* as opposed to one that has been slit apart, and being *first* in line to get a drink on a hot day. It is very important, therefore, to investigate these ideas through role play, drawings, and positioning a child in space to illustrate the topological notions, as well as by describing them verbally (see Reisman, 1977, for further discussion of transformational geometry and topological concepts).

Assessment Activities:

The topological notion of order is more complex than the other three. It may even be a lower-level generalization when it refers to ordinality of numbers, since the idea of sequence is generalized across several objects in a set. However, order in a spatial sense does not require the concept of number. Order is related to sequencing; there can be order of (a) events, (b) length, (c) size, (d) color, or (e) two or more sets, as shown here:

1.

a) Open a closed door and then walk through.

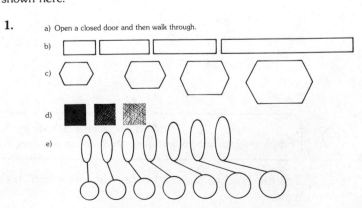

2. Give children experiences viewing a line of trees, for example, from the side (the road) and also from the front. This may be easier to accomplish by taking slides of trees in the neighborhood and showing them in class. Ask a child to point to the same tree in both views.

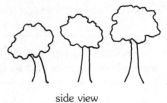

side view   front view

3. Have children help with bulletin board displays that develop order relationships. This may involve starting at one end and filling in toward the other.

Teaching Activities

Proximity, enclosure, separateness and order, even though basic for some, may not be assumed for the child who has difficulty attending to salient aspects of a situation. For such a child these relationships must be made apparent. Basic topological relationships should be within the grasp of *all* children who are capable of being educated in the mainstream. Of course, children will acquire them at different ages and at different speeds; some will need more repetition of activities to develop these relationships. The use of these topological relationships in forming generalizations will benefit students to varying degrees. The following instructional strategies focus on teaching the topological relationships of proximity, enclosure, separateness, and order.

Instructional Strategy 5.3.1:
Reinforce attention to a relevant dimension.

Using reinforcement that incorporates the relevant dimension enhances attention to that dimension (Fisher & Zeaman, 1973). Applying this to proximity, you might say, "I like the way you are noticing the children near you" or "I like the way you are standing near your desk."

Instructional Strategy 5.3.2:
Point out relevant relationship.

To help children construct the relationships of proximity, enclosure, and separateness it is important to isolate the relationships. A child may be unable to discover a particular relationship because he attends to unrelated aspects. Therefore, you must point out that he is near to an object and far away from another. For a retarded child, use a minimum of words—omit extraneous words in helping him to see these spatial relationships. For example:

    a. Mary near desk. (proximity)
    b. John inside tire. (enclosure)
    c. Puzzle apart. Put together. (separateness)

> *Instructional Strategy 5.3.3:*
> *Incorporate movement to develop order.*

Bodily movements are important in developing notions of order. If the child has been aware of rhythm, turning in space, and traversing various distances, he will have acquired basic experiences for understanding order relationships. Children who are developing at a slow cognitive rate may not attend to order relationships that are embedded in a situation, and so the teacher will have to point out the order. For example, a teacher of mentally retarded adolescents was trying to teach the game of badminton. He quickly discovered that saying "Throw the bird up into the air and hit it with the racket" was not enough structure for his students. They were not attending to the order of movements, and they achieved some success in the game only after the teacher helped each student to move through the sequence of events necessary to serve the bird.

5.4.0 Arbitrary Associations

This chapter also focuses on teaching arbitrary associations. Word names and digits (0, 1, 2, . . . 9) that represent numbers are examples of arbitrary associations in arithmetic. Word names and corresponding graphic symbols that designate regular polygons ("square," "rectangle," "circle") are examples of arbitrary associations in geometry. All of these examples involve essentially the same kind of learning, namely, discrimination learning. This type of learning task was described by Blake (1976).

Assessment
Activities: 1. Determine if the child uses number words as adjectives. Observe her speech to notice such phrases as:

            a. "one" mouth
            b. "no" candy
            c. "two" hands
            d. "ten" fingers

        2. Determine if the child uses number as a noun.

            a. I have "two."

    b. I have "none."
    c. I have "seven."

3. Determine if the child replaces the word expression with digits, that is, replaces the verbal by the use of graphic symbols.

    a. Child writes the digit for each of the following words: one (1), two (2), . . . nine (9), none or zero (0).

4. Determine if the child uses ordinal numbers. Observe his speech to notice such uses:

    a. I want to be first.
    b. He got here last.
    c. I'm second.

5. Present a flashcard with a digit and ask the child to say the digit's name.

    a. Show | 2 | , child says "two."
    b. Show | 5 | , child says "five."

6. Say a number name and ask child to select its digit from digit cards 0–9. Set out digit cards in random arrangement.

    a. Say "four"; child hands you the "4" card.
    b. Say "one"; child points to the "1" card.

Teaching Activities

> ### Instructional Strategy 5.4.1:
> *Emphasize differences in distinctive features of stimuli.*

Retarded students are weak in selecting and attending to relevant attributes of a situation—especially in discrimination learning (Zeaman & House, 1963). This suggests that a retarded child would benefit by writing digits if no *overlap* is present. Blake (1974, p. 202) pointed out that:

> Overlap among distinctive features between stimuli increases confusion and makes discrimination learning harder. . . . Overlap among distinctive features between different stimuli should be reduced. When overlap cannot be prevented, instructional procedures should include techniques to focus attention on the overlap or nullify its confusing effects. . . . The distinctive features for words are the letters/sounds which compose them. Consider the words *dog* and *pet* in relation to one another; each is characterized by a unique set of letters and sounds with no overlap . . . consider overlap among *too, two, to;* among *road, rode, rowed;* . . . with homonyms, we can reduce the overlap among distinctive features. However, we can focus attention on the overlap by various kinds of coding . . . color coding or simple underlining and pretraining.

Table 5.1 is a matrix of distinctive features of the digits 0–9. By looking across this matrix, it becomes apparent that two digits with no features in common are 0 and 2. These digits may be placed on an overhead projector to show their distinctive features. When differences and similarities in the shape of digits are compared, point out such features as vertical/horizontal lines, open/closed curves, and intersections of lines or curves. An example of transparencies illustrating that 0 and 2 have no overlapping characteristics is shown here:

As an activity for a gifted young student, have her list (a) digits with no overlap and (b) digits with overlap in symmetry, in closed-curve feature, in open-curve, etc. (for example, no-overlap digits: 0, 2; 0, 4; 0, 5; 0, 7; overlapping in symmetry: 0, 1, and 3; overlap in closed-curve feature: 0, 6, 8, 9; overlap in open-curve feature: 2, 3, 5, 7; overlap in intersection feature: 4, 6, 8, 9).

In 6 and 9, the distinctive feature is the position of the closed curve in relationship to the vertical line. In the digits 3 and 5, their direction is a relevant feature. These should be emphasized.

Other techniques regarding cue distinctiveness were summarized by Mercer and Snell (1977, p. 111) as follows:

The relevant dimensions of a stimulus need to be distinct in order to improve attention. Relevant dimensions of instructional stimuli should be highlighted. For

Table 5.1. *Distinctive Features of Digits*

Feature	Digit 0	1	2	3	4	5	6	7	8	9
Straight										
horizontal			+	+	+	+		+		
vertical		+			+	+				
diagonal								+	+	+
Curve										
closed	+						+		+	+
open			+	+		+		+		
Intersection					+		+		+	+
Redundancy										
cyclic change			+	+		+	+		+	+
symmetry	+	+		+					+	
Discontinuity										
vertical		+			+					
horizontal			+	+	+					

Note. Adapted from Eleanor J. Gibson, *Principles of perceptual learning and development*, ©1969, p. 88. Reprinted by permission of Prentice-Hall, Inc., Englewood Cliffs, New Jersey.

example, if a child is to respond to a picture of a person's face and depict the mood reflected in the picture, the features of the face which cue mood (mouth, smile) should be prominent. Cue distinctiveness is also enhanced by decreasing background cues, i.e., put stimulus on plain paper. Other techniques to promote cue distinctiveness may involve:

(1) using raised letters, pictures, shapes
(2) presenting only one stimulus at a time, i.e., one letter, one number, one face, one word, one sentence
(3) using a highlight pen to draw attention to relevant dimensions . . . using arrows to denote relevant dimensions . . . using colors . . . drawing circles around selected items.

Instructional Strategy 5.4.2:
Control irrelevant stimuli.

If the student is to attend to the shape of a geometric figure or to the number property of a set, then omit as many of the other attributes as possible. For example, omit unnecessary colors, textures, or sizes of objects used for instructional purposes. Select textbook pages that do not have a busy, distracting format, and remove extraneous materials from the immediate work area.

Instructional Strategy 5.4.3:
Use the student's preferred dimension.

Fisher and Zeaman (1973) found that students indicated preferred attributes and learned faster when such attributes were relevant to a learning task. Techniques for determining which attributes of a stimulus are preferred were presented by Mercer and Snell (1977, p. 111) as follows:

> The mentally retarded child's dimension bias may be informally tested and then featured on instructional stimuli. To test such a bias, like items may be presented which differ in size or color and the child is asked to select the item he prefers. For example, the teacher could present a series of triangle shapes which differ in color and size, record the student's responses, and determine which dimension is most frequently selected. . . . If a child has a bias to attend to color, each of the arithmetic signs (+, −, ÷, ×) could be paired to a color. Minus signs could always be red, plus signs blue, and so on. Gradually, the color could be faded out, leaving the child responding only to the sign.

Instructional Strategy 5.4.4:
Place stimuli to be discriminated in close proximity on initial tasks.

Increased distance between two stimuli makes discrimination learning more difficult for retarded children (Evans & Beedle, 1970). If the child is to discriminate one digit from another, for example, use close placement rather than distant.

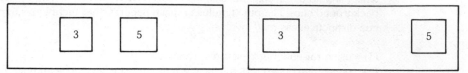

Instructional Strategy 5.4.5:
Increase the number of relevant dimensions.

Increasing the number of relevant dimensions of an item to be selected from two or three choices aids retarded students (Ullman & Routh, 1977). For example, if a child is to select the symbol 3 from a set, write the 3 in a different *color* and a different *size*.

Instructional Strategy 5.4.6:
Use cues that are different during initial learning.

Present shapes or numbers that are significantly different in appearance. For example, have the child discriminate between 2 and 6 rather than 9 and 6 during initial stages of recognition (Fisher & Zeaman, 1973).

Instructional Strategy 5.4.7:
Include identical stimuli to serve as background and thus to aid attention to relevant cues.

Brown's (1970) research suggests using a sequence of identical stimuli in a discrimination task for mentally retarded children. For example, if the child is to choose the digit 6 from a sequence of five, present the following:

7   7   7   6   7

Instructional Strategy 5.4.8:
Reinforce alternative positions on tasks involving random response patterns.

Mercer and Snell (1977) reviewed research on response strategies or biases of mentally retarded children. They suggested the following:

> On instructional tasks which require random response patterns, provide pre-training to retarded persons who have response sets. If a retarded child tends always to select objects to the left, pretraining which includes reward for attending to stimuli in other positions should help.   (p. 112)

For example, have the child select the digit 3 from each of the following choice sets. Notice how the position of the 3 is alternated.

7	3
3	5
4	3
3	1

Summary Thought Questions and Activities

I.   Discuss how the following summaries help you to provide instructional strategies in mathematics for a slow learner.

A.   Varying the color, size, or texture (that is, stimulus dimension) of a member of a sequence and organizing a sequence by various groupings (called "chunking") are particularly appropriate for learning sequences with little inherent meaning, such as digit sequences or the alphabet.

B.   Mentally retarded have difficulty in organizing material to be learned. This inhibits their retrieval processes (Spitz, 1973). Therefore, material should be grouped for presentation, rather than arranging it randomly. Suggestions for grouping materials include spacing, redundancy, and presenting stimuli simultaneously rather than sequentially.

C.   To facilitate recall, encourage children to use rehearsal strategies (such as verbal or imagery). In verbal rehearsal, the child labels components of a sequence either silently or aloud. In imagery rehearsal, the child associates the concept to be learned with an object, a picture, or a circumstance. This aids recall.

II.   How would you adapt the following instructional activities to meet the needs of children developing at different cognitive rates? Focus on helping them to construct basic relationships of the kind discussed in this chapter.

A.   practice activities
B.   pacing of presentation
C.   questioning
D.   response requirements
E.   rest/activity periods
F.   use of cueing techniques
G.   mode of presentation of mathematical ideas (concrete, picture, graphic symbols)

    H. mode of representing ideas during practices (concrete, picture, graphic symbols)

    I. modifying conceptual load of a lesson

III. List basic relationships that are prerequisite to the following concepts:

    A. recognizing shape and form
    B. determining sameness and difference
    C. learning quantity and/or size through comparison
    D. developing spatial orientations
    E. classifying
    F. sorting

IV. Select three students of the same age but who are developing at a slow, average, and rapid cognitive rate, respectively. Administer some of the instructional strategies suggested in this chapter, and list your observations of the children's performance on the various tasks. Keep in mind the cognitive generic factors that affect learning as you observe the children. The following characteristics, suggested by Krutetskii (1976), should be used as a guide for your observations:

    A. preferred hand
    B. coordination
    C. fatigue
    D. distractibility/short attention
    E. perseveration
    F. observe the child's speech and understanding of directions
    G. systematic approach vs. trial-and-error
    H. match color and shape
    I. reproduce simple designs
    J. work from left to right
    K. need for repetition of directions
    L. independence in completing each task
    M. process information slowly, rapidly, or erratically

Suggested Readings

Bricker, W. A., & Bricker, D. D. Receptive vocabulary as a factor in the discrimination performance of low-functioning children. *American Journal of Mental Deficiency,* 1971, 75, 599–605.

Engelmann, S., & Carnine, D. *Distar arithmetic: An international system.* Chicago: Science Research Associates, 1970.

Fernald, G. M. *Remedial techniques in basic school subjects.* New York: McGraw-Hill, 1943.

Flavell, J. H. *Developmental studies of mediated memory.* In H. W. Ress & L. P. Lipsett (Eds.), *Advances in child development and behavior* (Vol. 5). New York: Academic Press, 1970.

Sattler, J. M. *Assessment of children's intelligence.* Philadelphia: W. B. Saunders Company, 1974.

Spitz, H. H., Goettler, D. R., & Webreck, C. A. Effects of two types of redundancy on

References

Blake, K. A. *Teaching the retarded.* Englewood Cliffs, N.J.: 1974.

Blake, K. A. *The mentally retarded: An educational psychology.* Englewood Cliffs, N.J.: Prentice-Hall, Inc., 1976.

Brown, A. L. Subject and experimental variables in the oddity learning of normal and retarded children. *American Journal of Mental Deficiency,* 1970, *5,* 142–145.

Cronbach, L. J., & Snow, R. E. *Aptitudes and instructional methods: A handbook for research on interactions.* New York: Irvington Publishers, Inc., 1977.

Ellis, N. R. Memory processes in retardates and normals. In N. R. Ellis (Ed.), *International review of research in mental retardation* (Vol. 4). New York: Academic Press, 1970.

Evans, R. A., & Beedle, R. K. Discrimination learning in mentally retarded children as a function of irrelevant dimension variability. *American Journal of Mental Deficiency,* 1970, *74,* 568–573.

Fisher, M. A., & Zeaman, D. An attention-retention theory of retardate discrimination learning. In N. R. Ellis (Ed.), *International review of research in mental retardation* (Vol. 6). New York: Academic Press, 1973.

Krutetskii, V. A. [*The psychology of mathematical abilities in schoolchildren*] (J. Kilpatrick & I. Wirszup, eds., and J. Teller, trans.). Chicago: The University of Chicago Press, 1976.

Mercer, C. D., & Snell, M. E. *Learning theory research in mental retardation: Implications for teaching.* Columbus, Ohio: Charles E. Merrill Publishing Company, 1977.

Reisman, F. K. *Diagnostic teaching of elementary school mathematics: Method and content.* Skokie, Ill.: Rand McNally, 1977.

Sattler, J. M. *Assessment of children's intelligence.* Philadelphia: W. B. Saunders Company, 1974.

Spitz, H. H. Consolidating facts into the schematized learning and memory system of educable retardates. In N. R. Ellis (Ed.), *International review of research in mental retardation* (Vol. 6). New York: Academic Press, 1973.

Torrance, E. P. *Encouraging creativity in the classroom.* Dubuque, Iowa: William C. Brown Company Publishers, 1970.

Ullman, D. G., & Routh, D. K. Discrimination learning in mentally retarded and nonretarded children as a function of the number of relevant dimensions. *American Journal of Mental Deficiency,* 1971, *76,* 176–180.

Waugh, N. C., & Norman, D. A. Primary memory. *Psychological Review,* 1965, *72,* 89–104.

Zeaman, D., & House, B. J. An attentional theory of retardate discrimination learning. In N. R. Ellis (Ed.), *Handbook of mental deficiency.* New York: McGraw-Hill, 1963.

Endnotes

1. The Digit Span Test is a measure of attention and short-term memory that requires the capacity to retain elements having no logical relationship. The Assessment Activity involves the Digit Span Forward Test, featuring the repetition of digits in the same sequence as given. For some, this is indicative of associative abilities. Others create either an auditory or a visual image of the digits and merely read them off. This task can be reversed; in the Digit Span Backward, the sequence is to be repeated in reverse order. This task involves a transformation of order and goes beyond simple recall.

2. A distinction between two types of redundancy was described by Mercer and Snell (1977, p. 17) in their summary of research on short-term memory: "When digit-span information had 50% redundancy . . . the couplet type of redundancy (3, 3, 9, 9, 7, 7) was easier for mildly retarded and equal MA normals to discover and use as a RS [*rehearsal strategy*] than was repetition redundancy (3, 9, 7, 3, 9, 7)."

Instructional Strategies Based on Cognitive Needs: Lower-Level Generalizations

6 The ability to generalize from abstractions is a sign of mathematical competence. The usual interpretation of a generalization (also referred to as a principle or rule) is that two or more concepts (abstractions) are related in some way. However, some generalizations are less complex and indeed form the basis from which concepts emerge. For example, when the one-to-one relationship is generalized across two sets whose elements may be matched one-for-one, we have the generalization of *equivalent sets*. This can then be extended to matching three sets in a one-to-one relationship. At this level, the child need not be able to name, label, or otherwise identify the cardinal number property (threeness, fiveness, and so on) which is a concept. She merely needs to recognize that all of the sets under consideration "have the same number." This is an example of a lower-level generalization. The matching of sets in this manner is prerequisite to the concept of number.

6.1.0 Duality

We must structure our knowing in order to "filter, store, relate, and communicate data successfully" (Albarn & Smith, 1977, p. 18). One mode of generalizing is to *dichotomize* our environment. Daily activities allow children to use the word "one" as a single entity, although not necessarily as a counting number. Children quickly notice the "twoness" of their environment and distinguish one from two. Whenever we form a class within a universe, the universe is divided into two parts: what is in the class and what is not in the class. If the class is called A, its complement is –A, or "not A"; the dividing of the universe into +A and –A is a dichotomy. The notion of duality underlies sorting.

Assessment Activities:

1. Ask the child to point out the symmetry (duality) in his body. Even a child with physical anomalies may be aware of dualities such as two eyes, two feet, and so on.
2. Ask the child to tell about or draw duality in animals (fish have two eyes; birds have two eyes, two wings, two feet).
3. Have child discuss similarities and differences of things (dogs have two eyes and two ears, but four feet and one tail).

Teaching Activities

The lower-level generalization of duality involves *bilateral, line,* or *mirror* symmetry (if a mirror were stood by the line of symmetry, the mirror image would contain the other half of the shape). The line or axis of symmetry can be vertical, horizontal, or diagonal, so long as both sides of the figure are balanced on either side of the axis.

(IS)

> **Instructional Strategy 6.1.1:**
> *Employ familiarity to develop notion of duality.*

Use objects in the child's environment in school, his home, on the way to school. Encourage the child to notice the bilateral symmetry in furniture (beds, tables, chairs, bookcases), buildings, flowers, and other objects.

(IS)

> **Instructional Strategy 6.1.2:**
> *Point out relevant relationship.*

Playing with tiles or mosaics often leads to forming symmetrical patterns that should then be pointed out to the child.

Have child drop paint or ink onto paper. Fold the paper while the paint is still wet, and then point out the resulting duality in the paint blot. Have the child fold and cut paper and make decorations or masks that illustrate duality. Whenever possible, encourage the child to become aware of relevant relationships on his own, rather than pointing out the relationship.

(IS)

> **Instructional Strategy 6.1.3:**
> *Use redundancy and point out relevant dimension.*

Mercer and Snell (1977, p. 138) discussed the use of redundancy (adding more constant stimuli) in oddity-learning performances of retarded learners: "Instead of instructing the learner to select the odd stimulus from among three stimuli, Brown had the learner select the odd stimulus from among five stimuli (four identical and one odd)."

Have the child mark (or point to) the shape that is different (see Figure 6.1).

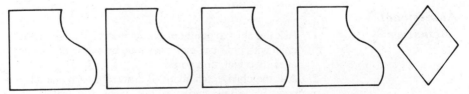

Figure 6.1. *Oddity-Learning Applied to Line Symmetry*

6.2.0 Equivalence of Sets

Sets having the same number of members are said to be equivalent. It is helpful to provide experiences that allow children to construct the notion of equivalent sets. Slow learners and retarded children can understand matching sets with two objects to other sets comprised of two objects. Matching is facilitated if the sets to be considered have some kind of a commonality and are comprised of objects that are familiar to the child—for example, cups to saucers; dolls to doll houses; and balls to games in which they are used. Matching colors to like colors or dominoes to dominoes is a more abstract task. Examples of matching equivalent sets of two objects and of three objects are shown in Figure 6.2.

*Assessment
Activities:*   1.   Match equivalent sets (see Figure 6.2).

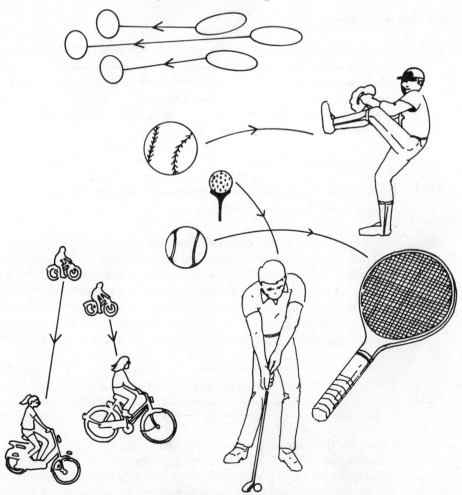

Figure 6.2.   *Matching Equivalent Sets Using Pictures*

2. Present equivalent sets by using concrete objects or pictures of objects. Ask the child to point to all sets with the same number of objects.

3. Show a set of two (three, five, . . .) objects to the child. Ask him to make another set that has the same number as yours. Have extra sets of objects available. These extra objects may consist of the following. Given a set with four red blocks to reconstruct, have available to the child (a) four red blocks identical to the model set, (b) four green blocks, (c) four assorted objects, (d) extra red blocks identical to the model set, (e) extra green blocks, and (f) extra assorted objects.

4. Extend activities 1 and 2 to creating *more than* only one set equivalent to the model set.

Teaching Activities

Comparison of sets is a prerequisite step of counting (enumerating). Equivalence of sets is not fully appreciated until the child understands comparisons involving *more* or *fewer*. The one-to-one relationship is basic to understanding the generalizations of equivalence, greater than, and less than. Following are instructional strategies in which the equivalence idea generalizes one-to-one correspondence to two or more sets.

> ### Instructional Strategy 6.2.1:
> ### Control the number of dimensions embedded in the stimuli.

For those children who become distracted by physical properties of objects, limit the number of these properties. For example, in the beginning activities use identical objects in all equivalent sets to be matched. Provide the child with a set of three pennies, for example, and ask him to make another set to match the original pennies. Then, as the child is able to perform this task, introduce an additional attribute that is irrelevant to constructing equivalence of sets. Thus, if red blocks were used, give the child blue or green or yellow blocks.

> ### Instructional Strategy 6.2.2:
> ### Use chunking to control distractions in the ground.

Using the previous activities, give the child objects one at a time for constructing the new set. Thus, in the penny activity, keep the extra pennies hidden in your hand and provide them one at a time as requested. Be careful not to cue the child when he has arrived at an equivalent condition. This avoids a figure-ground confusion for those students who have perceptual problems, and reduces the visual "noise."

> **Instructional Strategy 6.2.3:**
> **Prevent incorrect responses.**

Pointing out the correct response, by using various cues, prevents a child from practicing incorrect responses. These cues may include use of position, size, or shape. When a child places objects into holes in a form board (as in Section 5.2.0 Assessment Activities), the position, size, and shape are cues employed. The holes in the form board can be considered as one set and the objects that fit into the holes as another set, thereby ensuring the formation of two equivalent sets.

6.3.0 Greater Than–Less Than Matching

The inequality principles of greater/less than evolve from the one-to-one relationship and underlie seriation and transitivity. Activities involving greater and less than are similar to those for one-to-one correspondence, with one added condition: One set has more elements than the other. Comparisons may involve length, number of objects, weight, volume, mass, size, or area. Selecting objects that are familiar to the pupil may be less distracting than using objects that are novel. This enables the pupil to attend to the mathematical comparisons, rather than to irrelevant attributes. Of course, boredom from overuse must also be avoided when selecting materials.

Greater than–less than ideas are expressed in English by words with inflectional endings, namely, comparatives and superlatives. Since intellectually slower children do not learn well by incidental instruction (Singer, 1964), this mathematical language must be pointed out.

Assessment Activities:

1. Circle the shorter boy.

2. Circle the happiest face.

3. Draw a set greater than this:

Child draws:

4. Circle a set less than the given set.

Activities 1–3 may be translated to a concrete mode of representation. Notice that these activities involve comparing sets without counting or enumeration.

Teaching Activities

For most children, vocabulary relating to inequality of sets usually arises naturally during play. Such phrases as "more than," "fewer than," and "less than" are used along with the equality expression, "the same number as." These ideas develop as the result of incidental learning. However, some children do not learn incidentally.

(IS)

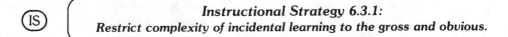

Instructional Strategy 6.3.1:
Restrict complexity of incidental learning to the gross and obvious.

Provide situations involving matching of sets where large differences in number occur. For example, have too few chairs at the table for the number of children who must sit; too few lunches one day so some have to wait for theirs; or too many tickets to a movie so they can invite some friends to go too.

(IS)

Instructional Strategy 6.3.2:
Encourage consistent responses to similar stimuli.

Retarded children show an inconsistency of response. When asking such a student to select the set with *more* (or *fewer*) objects, be specific and detailed in your directions, and correct anticipated errors by reminding the child of his previous correct selections during the one-to-one matching activities. If the child is able to match consistently most of the elements in a one-to-one relationship, then he can be encouraged to call the set with objects left over— with objects that cannot be matched—the larger set, or the set with "more."

> ## Instructional Strategy 6.3.3:
> *Replace incidental learning tasks with structured intentional learning tasks.*

Direct the child to touch the elements in a set that cannot be matched one-for-one to elements of another set. Direct her to then create another set that has more (or fewer) things than the set originally presented. During initial teaching of the greater than–less than extension of the one-to-one relationship, use sets that are logical and realistic—for example, playing musical chairs, in which there is one more child to be seated than chairs available when the music stops. The child who has no chair to sit on soon learns the rule of the game. Have the child who is left without a chair explain why he is standing. This will verify that he understands that there were more children than chairs.

Extend this game to the picture level. Show three pictures of boys in the class and two pictures of chairs (see Figure 6.3). Ask, "Is there the same number of chairs as boys?" If the child answers correctly (no), then move on and say, "Point to the set with more," or "Tell which set has more." Use the words "fewer" and "greater," emphasizing the suffix *er*, as well as "more" and "less" to point out use of comparatives to describe quantitative situations.

Figure 6.3

If Piaget is correct, children will not come to a full understanding of the one-to-one relationship or greater–less than ideas until they engage in concrete operational thinking. For example, if you have shown the child pictures of three chairs and pictures of three boys (one per picture) and if he has agreed that there is the same number of chairs as boys, spread out one of the sets and repeat the question, "Is there the same number of chairs as boys?" This task assesses whether the child can look at more than one aspect of a situation at a time. Can he apply the law of compensation? "The spread-out row takes up more space but no pictures were added, so there must still be the same number of pictures." This idea of nothing added, nothing changed involves a grasp of additive identity. Unless the child has developed these notions, related mathematics learning tasks will be learned in a rote manner with little understanding.

6.4.0 Sorting

Sorting, an extension of matching, is a perceptual task that is prerequisite to classifying. Pupils may sort things by external perceptual attributes such as shape, size, color, texture, or function. No abstracting (conceptualizing) is necessary in a sorting task.

A young child (or an older child who is developing at a slower cognitive rate relative to his age peers) notices only global, gross, obvious differences and similarities. In fact, when a young child is asked to describe how two objects are alike, he or she often will tell you how they are *different*; the implication is that noticing differences is easier. This makes sense, since pointing out similarities involves a thread of classifying, which in turn relies upon one's ability to abstract salient features of two or more situations.

It is very important to allow children to use as many senses as possible when investigating similarities and differences. By handling an object and seeing its uses, a child notices properties that are not initially apparent. Questioning by the teacher may serve as a cueing technique to facilitate a child's ability (a) to discriminate among some objects that are similar and (b) to tell ways in which they differ. Topological properties (discussed in Chapter 5) are helpful in pointing out differences in shape, size and position in space: Are things closed or open? Large or small? Near or far? Instructional strategies concerning how quickly a child learns may be summarized as follows:

- Adjust the amount of practice in handling objects.
- Adjust the type of cueing in pointing out differences and similarities.
- Select objects that are familiar and useful.

Assessment Activities:

Employ free choice or teacher modeling strategies in the following activities:

1. Sort by shape, placing similar objects in each of two boxes:
   a. Put round things (balls, beads) in one box and things with corners (block, box) in the other.
   b. Put things with holes (donut, flannel cut-out of zero) in one box and solid things (brick, block) in the other.
   c. Put intact rubber bands in one box and broken ones in the other.
2. Sort by size—large objects in one box, small things in the other.
3. Sort by color.

4. Place objects within an enclosure on the floor (hoola hoop or rope) in one box and those outside of the enclosure in the other box.

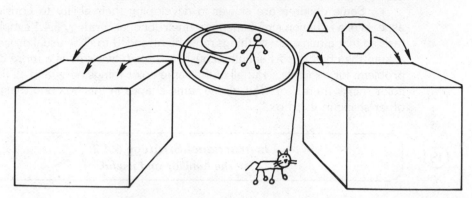

5. Sort by function: Place cooking utensils in one box and carpenter's tools in another.
6. Sort by likes and dislikes:
    a. Place pictures of food a child likes in one pile and foods she dislikes in another.
    b. Sort by favorite toys and those less preferred.

Teaching Activities

In the following tasks a matching by appearance is involved. The child may not as yet be able to abstract and label the attribute.

> ### Instructional Strategy 6.4.1:
> ### Reduce complexity of task.

Give the pupil one box, four round red blocks, and four red square blocks. Tell him to put all the same *shape* things in the box. He should do as shown in Figure 6.4.

Figure 6.4

Note, however, that if you were to state the instruction "Put all round things in the box," the task would no longer be a sorting task. The child would have to know what roundness is in order to group objects according to their

81

roundness. The distinction thus should be made between *sorting* and *classifying.*

Some children are slower in developing their ability to consider two aspects of a situation simultaneously. Instructional Strategy 6.4.1 simplifies the task so that a minimum of things must be attended to. The usual procedure of asking the child to sort simultaneously to obtain two sets is a more complex problem; for example, "Put all of the same color things here, and all the other color things there," or "Put all the same shapes in this box and things of the other shape in that box."

> **Instructional Strategy 6.4.2:**
> *Use the familiar and useful.*

Provide the child with objects that are familiar to her and preferably useful. The following examples may be sorted into two groups:

1. money: coins or paper bills
2. buttons: round or not round, two holes or four holes, plastic or wooden
3. clothes: hot weather or cold weather

> **Instructional Strategy 6.4.3:**
> *Use the strategy of becoming more familiar with the familiar.*

Use a child's creative strength by incorporating the following readiness activity into the sorting task. Prior to asking three or four children to sort a group of materials, engage them in the following activity described by Torrance (1970, p. 25):

> Select some common object in your environment . . . a rock, a piece of wood, a stick—almost anything. . . . Use every method you can think of to get to know this object without damaging it. Feel it, smell it, listen to it, experiment with it, imagine yourself in its place—whatever you can think of. Write down each [idea] that comes to you. . . . Do not stop too early! . . . Remember . . . , it does take effort!

Using Torrance's strategy, give each child an object to get to know better. Following this activity, ask each child to place the object that he or she had been "getting to know better" in a specified place (a shoebox on your desk). Then, in private, ask each child to sort the entire group of these objects. So long as your directions are the same for all children, you may specify or leave unspecified the number of groups into which the objects are to be sorted. You may either photograph this with an instant-developing camera or simply list objects comprising each group in the sort. Follow this up with each child showing his sorting organization to the others.

After the children have shared their sortings with one another, have them "reflect upon the experience." Torrance (1970, p. 25) suggests the following questions: "What ways did you use in getting to know your object? What resources [ideas, strengths, experiences] did you find that you had for knowing the object?" Ask how the object they got to know affected their sort.

You may use this activity with children at many levels of cognitive development. Even the slower-learning child may have creative potential, just as the child with high scores on achievement and IQ tests may be rigid and unoriginal in his or her thinking.[1]

6.5.0 Many-to-One and One-to-Many Matching

This is a generalizing of the one-to-one relationship to situations in which one set has a greater number of elements than another in some sort of matching pattern. The many-to-one generalization is an embryonic form of multiplication, which in turn underlies place value notation. The idea of ratio also derives from this many-to-one (or one-to-many) idea. Children developing at a slower cognitive rate will need more practice than others with many-to-one and one-to-many situations. Also, it is necessary to specifically point out this idea to those who have difficulty noticing salient attributes or conditions of a situation. For children who seem to process information more slowly, it is helpful to repeat activities in the same manner and using the same examples.

Assessment Activities:

1. Match five fingers to one hand, five toes to one foot.
2. Have children investigate petals on a flower, pine needles on a branch, children to a parent and parents to a child.
3. Given a set of coins (penny, nickel, dime, quarter), child is asked to match five pennies to a nickel, two nickels to a dime.
4. Discuss the idea of efficiency in reviewing the notching of a stick where one mark represents more than one of something.

Teaching Activities

Many-to-one matching is a component of many mathematical ideas. Yet it is a concept so obvious to adults that we are not always precise in pointing out this extension of the one-to-one relationship to children.

> *Instructional Strategy 6.5.1:*
> *Use consistent vocabulary.*

There are many ways of expressing a many-to-one or one-many idea. Usual phrases include:

1. "The notch on the stick *stands for* five sheep."
2. "Five pennies *equal* a nickel."

3. "Make a 10-for-1 *exchange*."
4. "Borrow a 10 and it *makes* 10 ones."
5. "A dime *is the same amount as* 10 pennies."
6. "One-half *is equivalent to* four-eighths."

Two basic ideas are expressed in the above phrases: representation and equivalence. The earliest enumeration systems used *representation*— notched sticks to represent groups of sheep, for example (see Reisman, 1977). Then, as a need for expressing greater quantity arose, the many-to-one idea was extended to mean "equivalent to." For example, the Egyptian symbol for a 10-to-1 idea was the heelbone (∩), while the 100-to-1 idea was expressed by a coiled rope (𝟗). The vocabulary to describe the use of these symbols should consistently state an equivalence idea, for example, "the heelbone is worth 10 strokes" or "the coiled rope equals 100 strokes."

Place *value* is also based on a many-to-one idea of equivalence. Instead of presenting a poor verbal model as in phrase 4, it would be better to consistently use language as in phrases 2, 5, or 6.

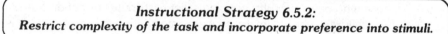

Instructional Strategy 6.5.2:
Restrict complexity of the task and incorporate preference into stimuli.

A "matching graph" (Dawes, 1977) allows children to match themselves or a symbol of themselves (such as their name card or their photograph) to their particular choice, as follows:

> from a discussion on favourite breakfast cereals, the teacher will put out on a table, or on the floor, empty cereal packers and the children will match themselves to their choice using lengths of coloured wool or tape which they will attach to the carton. . . . Again, so that they can all have a full view of the complete graph, they can replace themselves by their name card so that they can walk round the table or sit on the floor round the cartons and talk about what they have done.   (p. 69)

The many-to-one idea of representation is constructed as the children discover that several of them chose the same box as their favorite cereal. Other matching graphs may consist of choosing one's favorite sport, animal, flower, story, and so on (see Figure 6.5).

6.6.0 Seriation

Seriation involves generalizing (for example, the greater than/less than principle) across elements in a sequence. In arranging straws of different lengths in a short-to-tall progression, the "greater-than" rule governs placement of the straws. This example illustrates generalization of an asym-

Figure 6.5.   *Many-to-One on a Matching Graph (Favorite Story Animal)*

metric, transitive relationship. *Asymmetry* is expressed in the short-to-tall progression of the straws; *transitivity* is shown by the fact that a particular straw, if shorter than another, is also shorter than all others that are longer.

Assessment Activities:

1. Have the child arrange objects in a sequentially systematic manner according to a specified attribute (shades of color, size, shape, length, and so on).
2. Provide opportunities for the child to practice comparatives that express *asymmetric, transitive* relationships: This straw is next because it's *longer* than that one, but *shorter* than those.

3. "Cross out the figure that does not belong in the series" (See Figure 6.6).

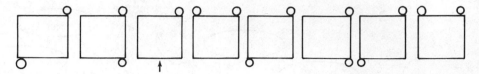

Figure 6.6

Note. From V.A. Krutetskii, *The psychology of mathematical abilities in school children* (J. Teller, trans., and J. Kilpatrick & I. Wirszup, eds.). Chicago: The University of Chicago Press, 1976, p. 152. Reprinted by permission.

4. "On the left are 4 or 5 figures that form a series. On the right are 5 figures. Find the one among them that should be the fifth (or sixth) one on the left" (see Figure 6.7).

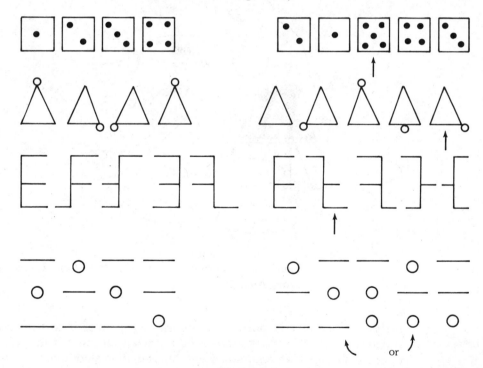

Figure 6.7

Note. From Krutetskii, 1976, p. 152. Reprinted by permission.

Teaching Activities

Some children cannot generalize a series because of perception factors involving differentiation and attention. They have not learned to discriminate

86

important features and do not pay attention to crucial relationships. Figure/ground characteristics of perception involve one's ability to select salient aspects of a situation. Thus, giving a child direction as to what to look for in a series will sometimes facilitate her constructing the necessary generalization. This adaptation to traditional figure/ground or embedded figures activities provides the action of *searching across a series*.

(IS)

> ### Instructional Strategy 6.6.1:
> #### Use a model whose competence in the task has been established.

Strichert (1974) found that retarded observers imitated competent models (competent with regard to performance of a task) more than noncompetent models.[2] He also found that noncompetent observers were more imitative than competent observers. This has implications for peer teaching: "although peer modeling is a viable tool, the competence of both the model and the observer must be considered" (Mercer & Snell, 1977, p. 230).

Successful performance on the assessment tasks may serve to identify "competent models," and these children may then be paired with slower-learning children. Assessment tasks 1, 2, and 3 for seriation (see pp. 85-86) may be used with peers serving as models.

Let us examine Assessment task 1 as an example of this instructional strategy. The peer model arranges objects in order by number of sides, from three sides to six sides (see Figure 6.8). There is no verbalization, but the peer model runs his finger along each edge to emphasize the number of sides. Then he says to his observer, "Now you put them in this same order." The model disarranges his seriation and says, "It's your turn. You do it." If the observer makes an error, the model immediately stops him (to prevent practice of incorrect response) and, if necessary, manually guides the observer's finger to run along the edges of the objects that had already been seriated. (This task is appropriate if the child cannot yet enumerate but is able to use a "one greater-than" rule.)

Figure 6.8. *Seriation Based on Number of Edges*

(IS)

> ### Instructional Strategy 6.6.2:
> #### Use of verbalization by model.

The competent model is seated next to a slower learner and states the principle by which she will seriate. "I will order these blocks [see Figure 6.9] by size. I put the smallest here, and the biggest over here [as she performs the task]. Then I find the one closest in size to the smallest. Then I put this one next —it's a little bigger, then the next bigger, . . ."

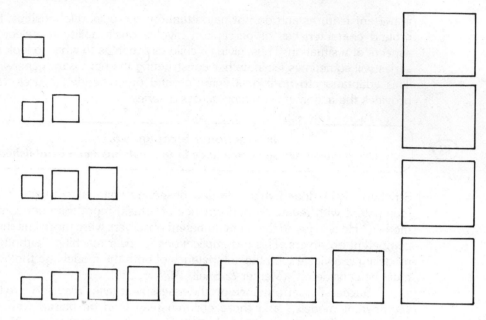

Figure 6.9.   *Seriation by Size*

(IS)   ***Instructional Strategy 6.6.3:***
Control the number of dimensions imbedded in the stimuli.

Have the child seriate by using dominoes, as shown in Figure 6.10. Again, as in IS 6.6.2, extension of a one-to-one relationship between dots on the dominoes to use of a "one-greater than" rule will allow the child to seriate by number of dots. The dominoes comprise only white dots on a black background and are otherwise identical, thus restricting the number of dimensions. This facilitates the child's focusing on the salient attribute for forming a series—the number of dots.

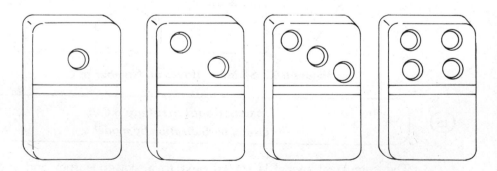

Figure 6.10.   *Controlled Number of Stimuli*

6.7.0 Judging and Estimating

These cognitive processes involve comparison and discrimination. At this level they are the components of thinking that permeate every type of decision. Judging and estimating are based on previous knowledge or experience; otherwise, the activity becomes guessing. Children who have difficulty with cause–effect relationships may have needed more direction with basic judgment and estimation activities.

**Assessment
Activities:**

1. Present simple problems in which a judgment must be made. For example, show the child a picture of a winter scene and ask, "Is it cold in this scene or is it hot?"
2. Extend task 1 to an estimation situation by asking, "How would you have to dress if you lived in this picture?"
3. Show the child the following attribute blocks.[3] Tell him to judge if the round shapes are all *larger* than the shapes with pointed edges.

4. Make use of estimation and judgment in real-life situations. Ask the children to describe how they estimate the amount of food to eat at one sitting.

Teaching Activities

In addition to the following suggestions, encourage parents to allow children to choose from among two choices at home: foods—hamburger or chicken for dinner; shirts—blue or striped; activities—movies or baseball game; and so on. Increase the complexity of judgments to be made according to the child's performance.

> **Instructional Strategy 6.7.1:**
> *Restrict complexity of the stimuli.*

Give the child one red block. Ask, "Is this red?" This question relies on the idea of duality discussed previously. A binary choice is involved—red or not red. Too often this basic level of judging is taken for granted. Some children need experiences in making binary judgments that are this simple to get them started.

> ### Instructional Strategy 6.7.2:
> *Provide sequencing experiences comprised of more abstract and/or subtle nature.*

A sequence extended ahead in time, involving an if–then condition, is more subtle and abstract in nature. Whereas judging deals with the present, estimation often involves a prediction. Estimating activities may derive from a judgment; for example, if you choose the striped shirt, what trousers will you select? Estimation of quotients in a division problem is a much more abstract activity. Those developing at a rapid cognitive rate still need these basic experiences.

> ### Instructional Strategy 6.7.3:
> *Encourage deferred judgment during problem-solving activity.*

Torrance (1970, p. 89) stated that "creative imagination and judgment cannot function concurrently at their highest level." This statement is in accord with the characteristic of creative thinking referred to as resistance to premature closure. It is suggested, therefore, that the nature of activity involving estimating and judging should vary.

A deliberate effort must be made to avoid too quick a decision. On the other hand, the student should have experience in developing criteria upon which a judgment may be based. For example, in planning a class store, many decisions must be made. What merchandise will be sold? Paperback books? Notebook paper? Pencils, pens? Book covers? Will the store be open to the entire school, to certain grade levels, or just to the class that runs the store? Who will handle the money? Where will the initial investment come from? At this stage, the students are generating questions that will help to identify salient problems.

In order to decide which problem or problems are to be solved, a judgment must be made. Perhaps the decision of which problem to attack first involves deciding on the purpose of the store. The teacher's purpose may be to provide students with real-life mathematics experiences; the students' purposes may be quite different. Thus, an awareness of one's own needs and motives comes into the picture.

Next, alternative solutions should be listed. These are then judged against some predetermined criteria, and a choice is made. Perhaps at this point a way is found to learn mathematics as the store is developed as well as to have a good time making plans for the store's opening. At every juncture along this problem-solving road, estimations and judgments were made at various levels of complexity. This kind of activity allows children at differing levels of cognitive development to engage in estimating and judging experiences.

90

Summary Thought Questions and Activities

I. Give specific examples of lower-level generalizations presented in this chapter.

II. Select a task involving seriation and list its prerequisites. For example, what must a child understand and be able to do in order to arrange pictures of people of different ages from youngest to oldest?

III. Explain the importance of including tasks that involve a simple binary choice for a child developing at a slower cognitive rate. Give an example.

IV. List reasons why children who are developing at an accelerated cognitive rate also need experiences with simple judging and estimating.

Suggested Readings

Blake, K. A. *The mentally retarded: An educational psychology.* Englewood Cliffs, New Jersey: Prentice-Hall, Inc., 1976.

Elkind, D., & Flavell, J. H. *Studies in cognitive development.* New York: Oxford Press, 1969.

Laurendeau, M., & Pinard, A. *Causal thinking in the child.* New York: International Universities Press, Inc., 1962.

Laurendeau, M., & Pinard, A. *The development of the concept of space in the child.* New York: International Universities Press, Inc., 1970.

References

Albarn, K., & Smith, J. M. *Diagram: The instrument of thought.* London: Thames and Heedson, Ltd., 1977.

Altman, R., & Talkington, L. W. Modeling: An alternative behavior modification approach for retardates. *Mental Retardation*, 1971, *9* (3), 20–23.

Arieti, S. *Creativity: The magic synthesis.* New York: Basic Books, Inc., Publishers, 1976.

Dawes, C. *Early maths.* London: Longman, 1977.

Krutetskii, V. A. [*The psychology of mathematical abilities in schoolchildren*] (J. Kilpatrick & I. Wirszup, eds., and J. Teller, trans.). Chicago: The University of Chicago Press, 1976.

Litrownik, A. J. Observational learning in retarded and normal children as a function of delay between observation and opportunity to perform. *Journal of Experimental Child Psychology*, 1972, *48*, 117–125.

Mercer, C. D., & Snell, M. E. *Learning theory research in mental retardation: Implications for teaching.* Columbus, Ohio: Charles E. Merrill Publishing Company, 1977.

Reisman, F. K. *Diagnostic teaching of elementary school mathematics: Method and content.* Skokie, Ill.: Rand McNally, 1977.

Ross, D. Effect on learning of psychological attainment to a film model. *American Journal of Mental Deficiency*, 1970, *74*, 701–707.

Singer, R. V. Incidental and intentional learning in retarded and normal children (Doctoral dissertation, Michigan State University, 1964). *Dissertation Abstracts International*, 1964, *25*, 652. (University Microfilms No. 64-7544)

Strichert, S. S. Effects of competence and nurturance on imitation of nonretarded peers by retarded adolescents. *American Journal of Mental Deficiency*, 1974, 78, 665–673.

Talkington, L. W., & Hall, S. M. Relative effects of response cost and reward on model on subsequent performance of EMR's. *The Journal of Developmental Disabilities*, 1975, 1(2), 23–27.

Torrance, E. *Encouraging creativity in the classroom*. Dubuque, Iowa: William C. Brown Company Publishers, 1970.

Endnotes

1. Arieti (1976, p. 342), in discussing creativity and intelligence, stated:

   Many researchers have investigated whether there is a correlation between intelligence and creativity. There is as yet no unanimous consent on the matter. The prevailing opinion is that highly intelligent persons are not necessarily creative.

2. Altman and Talkington (1971) discussed modeling and imitation in relationship to Skinnerian operant techniques. They proposed that modeling has the following advantages:

   1. Tasks need not be reduced to bits of behavior. To do so often leads to teaching splinter skills that cannot be generalized to real life or to academic tasks, as in extreme models of prescription teaching.
   2. The problem of identifying effective reinforcers is obviated, since reinforcement is not a required condition when one learns by observation. (Other researchers, however [Ross, 1970; Talkington & Hall, 1975; Litrownik, 1972], have found that associating rewards facilitates imitation by retarded observers.)
   3. Endurance for the behavior that was acquired or changed is longer for observation than for operant techniques.
   4. Learning is more generalizable in learning by observation than by operant procedures.

3. The attributes of these plastic blocks are as follows:
   a. Shape: △ □ ○ ⬡ ▯
   b. Color: red, yellow, blue
   c. Thickness: thick, thin

Instructional Strategies Based on Cognitive Needs: Concepts

7 What abilities are involved in constructing concepts? Let us take, as an example, the concept of *triangle*. First, the student must establish a one-to-one correspondence between perceptible properties of at least two representations of a triangle. He makes comparisons that lead to judgments of *similarity* and *difference*; he makes judgments in terms of both physical identity and perceived identity;[1] he can tolerate modification of less essential properties, while simultaneously responding to salient or essential properties; and he can group objects which are physically different—perceptually discriminable—according to some self-imposed rule or generalization. Vurpillot (1976, pp. 283–284) notes that:

> The class of equivalence brings together in the same category discriminable objects to which the subject gives the same response. Even when they are perceived as different, several objects are considered as equivalent because they have certain properties in common.

It is apparent from these prerequisites that the developmental mathematics curriculum is closely related to cognitive development. The concept of *triangle* is constructed from the child's ability to employ a one-to-one correspondence between features of two objects, and to focus sequentially on these features. Basic topological relationships are involved, especially *separateness* and *enclosure*. Also included in developing the concept of *triangle* is the ability to distinguish same from different and to make judgments concerning properties of objects.

Vurpillot (1976, pp. 281–282) discussed *perceptual invariance* as follows:

> Perceptual constancies provide an example of when identity of a particular object is maintained across modifications of its context. . . . When the form which has been learned is recognizable independently of its context and of modifications to all points other than those which define it, it can be described in terms of a set of relationships: a closed figure with three angles (or three sides). At this point it is no longer a matter of perceptual, but of conceptual invariance.

The crucial transition from perceptual invariance to conceptual invariance completes a major stage in the development of mathematical thinking and of cognitive development. The child can consistently attend to relevant properties, while at the same time discounting properties that are irrelevant to the perceived identity. The child is now aware of different examples of a class or category. Given different shapes of triangles (equilateral, isosceles, right, acute, obtuse), he will group them all as triangles, in comparison to circles, squares, rectangles, or some other simple, closed curve. If he is abstracting the concept *redness*, he will group all red objects apart from yellow or blue, ignoring differences in shades of red. If he is abstracting the number 3, he will select all groups with three members—regardless of the color, shape, size, texture, or function of the objects comprising the groups.

A concept is an abstraction—an idea. The concepts of *form* and *color* can be matched directly to concrete referents in the real world. The concept of *number*, however, does not have as direct a referent. It may be expressed by any object, group of objects, or lack of object (null set). The idea of number emerges from a synthesis of one's abilities to classify (that is, to consider a set on some attribute, such as color, size, age) and to seriate.

Assessment Activities:

1. Present two equivalent sets having elements with like qualities, such as two sets with three red wooden blocks in each set (see Figure 7.1). Say, "tell me what is the same about these two sets."

Figure 7.1

   (a) Both have red things.
   (b) Both have wooden things.
   (c) Both have blocks.
   (d) Both have the same spatial arrangement.
   (e) Both have things that feel the same.
   (f) Both have things that are of the same shape.
   (g) Both have three things.
   (h) Both have things that are the same size.

2. Add a third set comprised of three green wooden blocks, as shown in Figure 7.2. Say "Tell me what is the same about these three sets."

Figure 7.2

Note that conditions *a*, *d*, and *h* were knocked out. For a very slow child, change only one attribute per activity. Allow enough experience in each activity before continuing. For the next activity, you may wish to allow the child to point to what he wants to change in a set. The last attribute left will be the concept (such as color, geometric shape, or cardinality).

3. Add a fourth set comprised of a small blue wooden triangle, a large white piece of paper, and a marble (see Figure 7.3).

Figure 7.3

Note that conditions *a–f* and *h* were knocked out. The only attribute that is common to all of the sets is that each contains three objects.

4. For a more complex activity, set up a situation (for example, using blocks) in which an aggregate of attributes forms a classification group. This utilizes more than one concept within a class. Groups are formed based on color, size, shape, and thickness.

5. Modify Activities 1, 2, and 3 by using different objects. Other properties that you may wish to abstract include: length, function, sound, and weight.

Teaching Activities

In order to carry out tasks of classifying, the student must make comparisons. Comparisons involve construction of identity and of equivalence and difference, and ability to make decisions based on criteria that are appropriate. When presented with a problem, the student must determine the salient relationships upon which to construct the concept.

Various organizational strategies for input of data have been studied (Spitz, 1963, 1966, 1973). The instructional strategies that follow are suggestions for implementing some of these *organization of input* procedures.

Instructional Strategy 7.1.1:
Use familiar categories to insure meaning for the student.

Initial stages of conceptualizing should be restricted to sorting objects into familiar groups. Have children place pictures of objects into two groups. Say, "Make two groups of things that are similar." The following list of objects may be used:

1. dogs, cats
2. horses, cows, flowers
3. spoon, cup, plate, meat, vegetables, bread
4. spoon, cup, plate, tree, flower, bush

Some children will be able to label their resulting sets; others will be able only to show you. Some may only be able to classify sets 1 and 2.

(IS)

> **Instructional Strategy 7.1.2:**
> **Provide bimodal input.**

Encourage the student to handle, see, and talk about an object's properties. This will facilitate his development of abstraction and classification skills.

(IS)

> **Instructional Strategy 7.1.3:**
> **Group input by category.**

Verbally present the student with a category name, and instruct her to list as many category members as she can in *1 minute.* For example, say, "Toys," and she is to name as many toys as she can in 1 minute. This activity may be extended by using other categories such as colors, forms or shapes, materials, numbers, children's names, girls' names. Of course, this activity may also be of a *free recall* nature—that is, unstructured or noncued, as in, "Say some names of things."

(IS)

> **Instructional Strategy 7.1.4:**
> **Increase task complexity.**

Use of logic or attribute blocks presents a challenging series of conceptualizations for children with accelerated cognitive development. An example of an attribute block activity would be the following: Using blue, red, and yellow triangles of two thicknesses, say, "Make groups such that something is the same about all the things within a group." (See Figure 7.4.)
Then change the objects. Introducing different properties described provides new categories.

(IS)

> **Instructional Strategy 7.1.5:**
> **Employ creative problem-solving strategy to facilitate concept formation.**

Vygotskii (1962) used a method (described by Hanfmann & Kasanin, 1942) of inducing artificial concepts as part of a problem-solving process incorporating both nonsense words and perceptual material.[2]
Torrance (1970) presented five steps for disciplined teaching of creative problem solving. These steps are applied to abstracting the concept *red.*
Step 1. Sensing problems and challenges. A box of prizes is positioned on the other side of a "magic line." A wizard sits on the line and allows only persons with a special ticket to cross over to the magic box to be given a prize. The wizard decides ahead of time what magic ticket (in the form of an

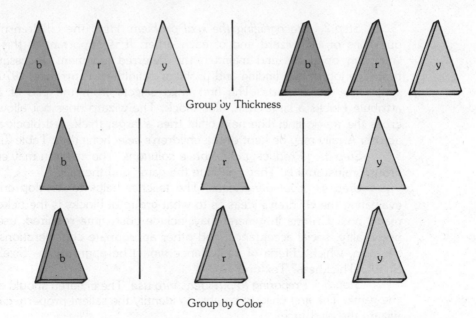

Group by Thickness

Group by Color

Figure 7.4. *Classification to Show the Concepts of Thickness and Color*

attribute block) he will accept, but he tells only the teacher (or some other verifier). See Figure 7.5.

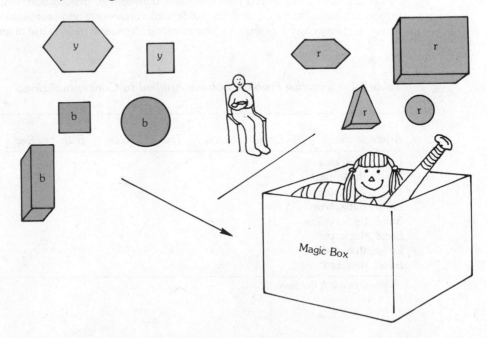

Magic Box

Figure 7.5. *Using Creative Problem Solving to Identify Concepts*

Step 2. Recognizing the real problem. Here the children must ask questions of the wizard and of each other. It is important at this step to "maintain openness and maintain the deferred judgment necessary for a thorough job of fact finding and problem definition" (Torrance, 1970, p. 80).

The game starts. The first child selects from the pile of assorted attribute blocks a large, thin, blue block. The wizard does not allow him to cross the magic line. The next child tries a large, thick, red block and gets across. A tally may be kept of the children's selections (See Table 7.1).

Step 3. Producing alternative solutions. The children may engage in group brainstorming. They continue the game and the tally.

Step 4. Evaluating ideas. The teacher helps to develop criteria for evaluating the children's ideas as to what group of blocks is the ticket to the magic box. Criteria in general may include cost, time required, usefulness, practicality, social acceptance, and other appropriate considerations. In this situation, which criteria of equivalence would be appropriate—Size? Color? Shape? Thickness? Texture?

Step 5. Preparing to put ideas into use. The children should complete the game. The first child to be able to identify the salient property can be the wizard the next time.

Torrance (1970, p. 83) listed skills necessary for the creative problem-solving process:

> making accurate observation, becoming sensitively aware of surroundings, fully utilizing all senses, shifting points of view, questioning, making both deliberate and random association, making predictions, organizing and reorganizing patterns, taking careful inventories, manipulating ideas, and making use of analogy.

Table 7.1. *Creative Problem Solving Applied to Conceptualizing*

Attribute Block	Large	Small	Tally Thin	Thick	Blue	Red	Yellow
Large, thin, blue	–		–		–		
Large, thick, red[a]	+			+		+	
Large, thick, yellow	–			–			–
Large, thick, blue	–			–	–		
Small, thick, yellow		–		–			–
Small, thick, red[a]		+		+		+	
Large, thin, red[a]	+		+			+	
Small, thin, red[a]		+	+			+	

[a] Allows crossing the magic line.

Summary Thought Questions and Activities

I.   List abilities involved in constructing concepts.
II. Describe the differences between perceptual and conceptual invariance.
III. Tell how the idea of number emerges.
IV. Discuss how the use of bimodal input enhances abstraction and classification.

Suggested Readings

Arieti, S. *Creativity: The magic synthesis.* New York: Basic Books, 1976.

Blake, K. A. *Teaching the retarded.* Englewood Cliffs, New Jersey: Prentice-Hall, Inc., 1974.

Blake, K. A. *The mentally retarded: An educational psychology.* Englewood Cliffs, New Jersey: Prentice-Hall, Inc., 1976.

Burton, T. *The trainable mentally retarded.* Columbus, Ohio: Charles E. Merrill Company, 1976.

Gruber, H. E., & Voneche, J. *The essential Piaget.* New York: Basic Books, 1977.

Krutetskii, V. A. [*The psychology of mathematical abilities in schoolchildren*] (J. Kilpatrick & I. Wirszup, eds., and J. Teller, trans.). Chicago: The University of Chicago Press, 1976.

Lovell, K. *The growth of basic mathematical and scientific concepts in children.* London: University of London Press, 1961.

Mercer, C. D., & Snell, M. E. *Learning theory research in mental retardation: Implications for teaching.* Columbus, Ohio: Charles E. Merrill Publishing Company, 1977.

Reynolds, M. C., & Birch, J. W. *Teaching exceptional children in all America's schools: A first course for teachers and principals.* Reston, Virginia: The Council for Exceptional Children, 1977.

Sattler, J. M. *Assessment of children's intelligence.* Philadelphia: W. B. Saunders Company, 1974.

Torrance, E. P., & Meyers, R. E. *Creative learning and teaching.* New York: Dodd, Mead & Company, 1970.

References

Hanfmann, E., & Kasanin, J. Conceptual thinking in schizophrenia. *Nervous and Mental Diseases Monograph,* 1942, *67,* 9–10.

Spitz, H. H. Field theory in mental deficiency. In N. R. Ellis (Ed.), *Handbook of mental deficiency: Psychological theory and research.* New York: McGraw-Hill, 1963.

Spitz, H. H. The role of input organization in the learning and memory of mental retardates. In N. R. Ellis (Ed.), *International review of research in mental retardation* (Vol. 2). New York: Academic Press, 1966.

Spitz, H. H. Consolidating facts in the schematized learning and memory system of educable retardates. In N. R. Ellis (Ed.), *International review of research in mental retardation* (Vol. 6). New York: Academic Press, 1972.

Torrance, E. P. *Encouraging creativity in the classroom.* Dubuque, Iowa: William C. Brown Company Publishers, 1970.

Vurpillot, E. *The visual world of the child.* London: George Allen and Unwin Ltd., 1976.

Vygotskii, L. S. *Thought and language.* Cambridge, Mass.: MIT Press, 1962.

Endnotes

1. Vurpillot (1976) distinguished between physical identity and perceived identity. In *physical identity,* complete congruence is involved in terms of an object's intrinsic characteristics: "the two objects have the same values on all possible descriptive dimensions" (p. 279). The *perceived identity* "always derives from the conservation of an invariance despite the existence of certain differences; the invariant differs only in terms of the type of identity which is sought" (p. 279). Thus, if the color red is the identity being sought between two or more objects, then the objects comprising each set must be of that color but can differ on other properties. If the number 2 is the identity, then all of the intrinsic physical properties of the objects in each set—size, form, color, or texture—are irrelevant. The relevant dimension for establishing identity in this case is the number property of each set, and this may be represented by any two discrete objects per set. "A true acquisition of the identity relationship has developed when the child can combine the idea of *same* in terms of the commonality of the properties and parts with an absence of differences" (Vurpillot, 1976, p. 311).

2. Hanfmann and Kasanin (1942) presented a very complex form of the concept formation task:

   The material used . . . consists of 22 wooden blocks varying in color, shape, height, and size. There are 5 different colors, 6 different shapes, 2 heights (the tall blocks and the flat blocks), and 2 sizes of the horizontal surface (large and small). On the underside of each figure, which is not seen by the subject, is written one of the four nonsense words: *lag, bik, mur, cev.* Regardless of color or shape, *lag* is written on all tall large figures, *bik* on all flat large figures, *mur* on the tall small ones, and *cev* on the flat small ones. At the beginning of the experiment all blocks, well mixed as to color, size and shape, are scattered on a table in front of the subject. . . . The examiner turns up one of the blocks (the "sample"), shows and reads its name to the subject, and asks him to pick out all the blocks which he thinks might belong to the same kind. After the subject has done so . . . the examiner turns up one of the "wrongly" selected blocks, shows that this is a block of a different kind, and encourages the subject to continue trying. After each new attempt another of the wrongly placed blocks is turned up. As the number of the turned blocks increases, the subject by degrees obtains a basis for discovering to which characteristics of the blocks the nonsense words refer. As soon as he makes this discovery the . . . words . . . come to stand for definite kinds of objects (e.g., *lag* for large tall blocks, *bik* for large flat ones), and new concepts for which the language provides no names are thus built up. The subject is then able to complete the task of separating the four kinds of blocks indicated by the nonsense words. Thus the use of concepts has a definite functional value for the performance required by this test. Whether the subject actually uses conceptual thinking in trying to solve the problem . . . can be inferred from the nature of the groups he builds and from his procedure in building them: Nearly every step in his reasoning is reflected in his manipulations of the blocks. The first attack on the problem; the handling of the sample; the response to correction; the finding of the solution—all these stages of the experiment provide data that can serve as indicators of the subject's level of thinking. (pp. 9–10)

Instructional Strategies Based on Cognitive Needs: Higher-Level Relationships

8

The next group of topics in the developmental mathematics curriculum (see Figure 1.4 in Chapter 1) involves transformations, set operations and relationships, number operations and relationships, and mathematical language.

8.1.0 Transformations

The mathematical term *transformation* reflects one's power to associate any idea or thing with any other idea or thing, no matter how similar or how different the two may be. Keyser (1922) used *Webster's Dictionary* as an example of transformation. The word entries are transformed into corresponding meanings, and each meaning into a word or words (synonyms). These transformations are built upon one-to-many or many-to-one correspondences. The *Dictionary* does not represent one-to-one correspondence, since a particular word may have several meanings, and synonyms are words with the same meaning.

Transformations occur when elements of one set are changed into those of another—for example, when a class of objects being counted is transformed into the class of real number pairs, or vice versa. Keyser (1922, p. 162) pointed out that analytical geometry "owes its very existence to a transformation . . . of points into number pairs and of such pairs into points."

There is another term that is closely related to the idea of transformation, and that is *invariance*. Transformations involve change. Invariant properties are those that remain the same under certain transformations. For example, if a circle is drawn on a rubber sheet and an X is marked within, no change in position between the X and the points on the circle will occur if the sheet is pulled or stretched, so long as the rubber is not torn. If a paper weight is slid across a desk, its weight, size, and shape will remain invariant—only its position in space will change. If five raisins are rearranged in space, the number property or cardinality of the set will remain the same. If a quantity of clay is transformed in shape, its weight and mass will not change. Some children are able to make such associations quickly and in great number; others take a long time to "see" these associations, and can make only a few.

Conservation

Related to transformation and invariance are the *Piagetian conservation tasks.* Conservation is the condition in which certain attributes of a class remain invariant (unchanged) under certain transformations. Some of these conservation tasks (which Piaget used to assess the stage of a child's cognitive development) include conservation of number, mass, and time. The tasks selected for Assessment Activities sample a child's ability to coordinate a constant with an accompanying change. This is characteristic of concrete operational thinking, which develops in most children from ages 7 to 11. This age range is normative. Extreme deviations in age of conserving provide important assessment data. The child of 5 who consistently conserves on these tasks is probably developing at a faster cognitive rate than the majority of children. The child of 13 who does not conserve—who states that "there are more raisins in the stretched out row"—is developmentally slower.

Assessment Activities:

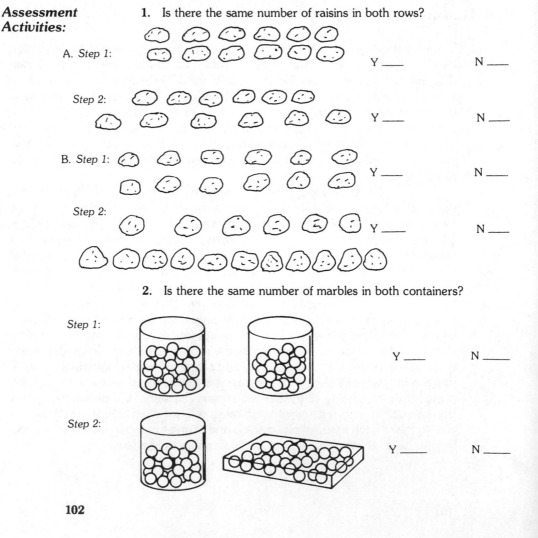

1. Is there the same number of raisins in both rows?

A. *Step 1:*   Y ___   N ___

Step 2:   Y ___   N ___

B. *Step 1:*   Y ___   N ___

Step 2:   Y ___   N ___

2. Is there the same number of marbles in both containers?

Step 1:   Y ___   N ___

Step 2:   Y ___   N ___

3. Is there the same amount of clay in the ball as in the other shape of clay?

A.

Y ___   N ___

B.

Y ___   N ___

4. Is there the same amount of clay in the ball as in the pieces of clay that were made from the ball?

A.

Y _____ N _____

B.

Y _____ N _____

5. Do both cars stop at the same time?

Starting positions   Same speed/Same time   Finishing positions

A.

Y ___ N ___

Starting positions   Same speed/Different times   Finishing positions

B.

Y ___ N ___

Starting positions   Different speeds/Same time   Finishing positions

C.

Y ___ N ___

Let us examine the notion of conservation more closely. What constructs underlie conservation? Piaget (1968) associates two ideas with conservation: equilibration and reversibility. Equilibration is an organism's attempt to reach a state of balance between opposite forces. Reversibility refers to the idea of inverses or reciprocals.

A clarification of the term *reversibility* is basic to (a) understanding how children learn mathematics, (b) selecting curriculum that is "learnable" by children in various stages of cognitive development, and (c) using instructional strategies that enhance mathematics learning. Somehow, the idea of reversibility of thought has pervaded English translations of Piaget's works. However, there is no such phenomenon as reversibility of thought in the context of learning mathematics ideas: "a physical or mental movement is never wholly reversible, since it occurs in time, and time is not reversible" (Piaget, 1965). In the digit span subtest of the Wechsler Intelligence Test for Children and in the Stanford-Binet, for example, repeating digits in reverse order does not involve reversal of thought. Examinees employ two main strategies for performing this task. Some use a visual array and simply read off the digits in right–left direction. Those employing an auditory array rehearse the sequence in a forward direction and simply call out the last digit in the sequence, working backward in this manner until they have spoken the first digit of the given series. The thought process is not reversed in either approach, but rather a successive sequence of mental activity ensues: "Expectation, intention, anticipation, premonition, and presentiment—all these have a forward reference in time" (Cohen, 1966).

In the conservation tasks, the child strives for balance (equilibration) between logic and perception. She accomplishes this by a series of successive approximations of relationships between change and no change—between a transformation and its simultaneous invariant condition. It is *the ability to reconcile simultaneous opposites* that underlies the notion of conservation.[1]

This interpretation of conservation allows the teacher to focus upon what a child *can* do, rather than on the observation that he is "not a conserver." Such an observation is no more helpful than saying, "He's below grade level." (If the child is a slow or retarded learner, of course he will be below grade level.) Neither of these answers tells what cognitive strengths the child has. The following questions elicit information that is helpful for formulating hypotheses about a child's cognitive strengths (see Reisman, 1978):

- Does she notice salient attributes of a situation? Does she ignore irrelevant information?
- Does she attend to more than one aspect of a situation at a time?
- Does she apply the mathematical law of compensation?
- Does she notice situations in which transformations destroy invariant conditions, as well as those in which invariants are maintained under transformations?
- Does she maintain some invariants and not others (for example, notices that number and mass remain unchanged under certain transformations, but not length or volume or time durations)?
- Is the child aware of her thinking process, or is she perception-bound?

104

- Is she able to make hypotheses and inferences?
- How accurate are her judgments and estimations?

Because the Piagetian tasks show what a child *is* performing, not what he *can* do, it is suggested that Piaget's protocol be used as an assessment screening test to obtain a general developmental level. The next step is to obtain answers to the questions listed previously in order to analyze the child's cognitive capacities.

Teaching Activities

Following are instructional strategies for investigating transformation-invariant relationships. The first instructional strategy provides a protocol for obtaining data on cognitive strengths.

Instructional Strategy 8.1.1:
Use modified conservation tasks.

The conservation tasks have been modified so that the child is given the invariant and the examiner (teacher) asks for transformations, instead of the other way around (as in the traditional Piagetian protocol). The child is encouraged to identify alternate transformations, to test them for the invariant condition, and then to evaluate whether or not they are allowed under the particular invariant. Given the invariant *number* or *cardinality* of a set, what transformations are allowable under the particular invariant? Can you re-arrange the elements of the set in space? Can you perform a one-to-one substitution of color? Size? Shape? Texture? Weight? Can you add elements? Subtract elements? Can you divide each element into parts, leading to a multiplication of parts? Can you go from discontinuous to continuous form (for example, melt six candy kisses into one blob of chocolate)?

Table 8.1 serves as a guide for developing a problem-solving instructional program by modifying the Piagetian conservation tasks. This procedure allows the teacher to observe a child's cognitive activity in terms of what the child is able to do. Following are suggested procedures for each of the invariants listed in Table 8.1:

Number: Give the child 6 pieces of candy. Say, "Tell me all of the changes that you can make but still keep six things in the group."
You may use some or all of the previous questions (see pages 104–105) to help the child get started.
This activity may be done more concretely for the child who is not able to think in a representational manner. For such a child, provide alternative objects that may be used for testing various transformations—for example, objects of different color, shape, size, texture, or weight.

Mass: Give the child a ball of clay. Say, "Tell me all of the changes you may make on this clay but still keep the amount the same."
You may provide a rock or rubber ball that is the same size (has the same mass) as the ball of clay, or different-color clay, and ask the child to show the allowable transformations.

Weight: Give the child a pound of jelly beans. Say, "Tell me all of the changes you can make but still keep the weight the same."
Provide a balance and objects of similar and different weights. Encourage the child to experiment with the possible substitutions while maintaining the pound weight.

Length: Give the child two rulers. Say, "Tell me the changes you could make with these two rulers and still have two objects with the same length."
Provide rulers of different color than the model, with straight edges of different width, material, thickness, or weight. Include two lengths that are made out of paper that can be cut into segments and then placed end to end to recreate the original length.

Area: Give the child a piece of paper. Say, "Tell me all of the ways you can change what I have given you and keep the size [area] the same."
Provide different-color paper, different material, different weight, different shape, but the same area, and encourage the child to experiment.

Volume: Give the child a glass of water. Say, "Tell me all of the changes you can make and still keep the amount of liquid the same."
Provide different-color water, milk, tea, glasses of different size and shape, sand, and ice cubes equal in volume to a given quantity of water.

(IS)

Instructional Strategy 8.1.2:
Use of peer team learning.

Investigating transformations and invariants lends itself well to peer learning. Children may be teamed in various combinations of performance level (high-high, high-low, low-low) and encouraged to engage cooperatively in activities described in Instructional Strategy 8.1.1.

(IS)

Instructional Strategy 8.1.3:
Apply modified conservation tasks to rigid transformations.

Translations (slides), rotations (turns), and reflections (flips) are dealt with in Table 8.1 under the column entitled "Rearrange in space." See Appendix A to test your knowledge of these relationships. The invariant in each is the relationship among the physical properties of an object; the transformations are *spatial*. For example, a book may be slid along a table, rotated in space, and flipped over, without changing any of its physical properties. The identity

106

TABLE 8.1. *Summary of Transformations Possible Under Selected Invariants*

Transformations											
	Rearrange in Space	Substitution						Arithmetic Operation			
Invariant		Color	Size	Shape	Texture	Weight	Add	Subtract	Divide Each Element	Multiplication Resulting From Division	Go From Discontinuous to Continuous Form
Number (6 candies)	+	+	+	+	+	+					6 chocolates melted into 1 blob
Mass (clay)	+	+	+	+	+	+		+		+	
Weight (jelly beans)	+	+	+	+	+				+	+	
Length (2 rulers)	+	+	+		+				+	+	+ Combine segments
Area	+	+		+					+	+	+ Combine parts of a plane equal in area to the given part
Volume (liquid)	+	+		+					+	+	+ Pour small glasses of water into one of equal volume to the sum of the small glasses

The "+" sign implies an allowable transformation that maintains the invariant.

of the book has been maintained.

> ### Instructional Strategy 8.1.4:
> ### Employ creative problem solving.

Give the child three blue blocks. The modified conservation tasks described in IS 8.1.1 involve the definition of conservation as one's awareness of simultaneous opposites. This allows one to resolve perceptual-logical conflicts through use of skills in creative problem solving. Creative problem solving employs both convergent and divergent thinking. The modified conservation tasks are both divergent and convergent, whereas Piaget's protocol is convergent.

Figure 8.1 illustrates the convergent-divergent cycle in the modified tasks. Convergent thinking occurs at points A, C, and E; divergent thinking, at B, D, and F. The points of the cycle are further explained:

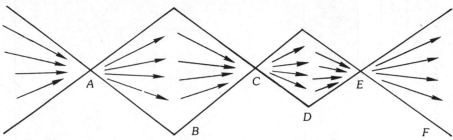

A. "Here is a set of three objects. Make as many changes in the set as you can but keep the number of things 3." Teacher identifies the problem. The question is concrete. The abstraction is given. The task is perceptual—the changes involve perceptual attributes.
B. Brainstorming occurs: "I can change color, shape, size, spatial arrangement, texture."
C. Judgment is made by subject as to correctness of his responses: "I can't add, or subtract, or break them up, or grind them into sawdust."
D. More responses emerge: "I can change the function—use three cars instead."
E. The subject judges new solutions.
F. The problem stays open; the child goes away still thinking about it.

Figure 8.1

A. The examiner identifies the problem, which is also the case in the Piagetian protocol. However, in the Piagetian tasks the question asks for the invariant (Is the number [length, volume, weight] the same?) and shows the transformation. Thus, in the Piagetian tasks, the abstraction—which is not readily perceived—is required, while the change—which is readily perceived—is given. In the modified conservation tasks, the opposite occurs.

B. This is the point at which the difference between the Piagetian and Reisman tasks is greatest. Divergent thinking is now elicited. The subject brainstorms:

"I can change them into a row."   (spatial arrangement)
"I can change to red blocks."   (color)

"I can change to circles."   (shape)
"I can substitute raisins."   (texture, function, color, shape)
"I can substitute toy cars."   (function)

C. After the brainstorming at point *B*, where most come up with perceptual or functional substitutions, a period of judging, evaluating, and estimating occurs. Transformations that are not allowable under the given invariant are stated:

"I can't add anything, or take away."
"I have to make changes one-for-one."
"I can't break things up or I'll have more."
"I can't crunch the blocks into sawdust, or I won't still have three things."

D. More responses may emerge that are allowable under the invariant. Some children may give other examples of a transformation already given, such as "I can substitute green blocks, or toy boats, or spread them out." Others may continue to test transformations resulting from mathematical operations (add, subtract, divide).

E. Again, the thinking becomes convergent as new alternative solutions are judged.

F. This problem-solving strategy tends to prevent premature closure. Subjects often talk about possible solutions for days afterward. Also, they are receptive to testing transformations allowable under other invariants (see Table 8.1).

The modified tasks develop creative problem-solving skills. They can be engaged in at the concrete level throughout, or by a combination of concrete-hypothetical, or hypothetical-hypothetical. *Concrete throughout* tasks would include concrete examples of alternative solutions (for example, have available objects of different color, size, texture, shape, and so on). The subject can then test solutions by manipulating objects present. *Concrete-hypothetical* tasks could include providing a few of the alternative solutions concretely, or merely showing the original problem concretely (presenting three blue blocks) and having the subject hypothesize additional transformations. *Hypothetical-hypothetical tasks* would show no concrete exemplars, but would involve describing the set verbally and discussing possible transformations.

8.2.0 Set Operations and Relationships

There are children whose rate of intellectual growth limits their ability to (a) deal with complexity, (b) abstract, (c) hypothesize, and (d) make judgments and estimations. They need to learn the mathematics that is most related to their life needs. Some of these needs involve higher-level relationships as well as higher-level generalizations (such as basic number facts, measurement, money value, budgeting, cause–effect).

To date, we do not know the breadth of mathematics that a mildly retarded child can learn. Placement in regular mathematics classes and interaction with faster learners, effects of mainstreaming on a child's perception of self, emphasis on developing creative problem-solving skills, enrichment of instruction for the slower learner as well as for the very rapid learner, and the use of calculators as accommodation to poor memory all are open-ended issues.

A child constructs basic relationships and lower-level generalizations that enable him to form classes or concepts. As the child plays with different objects, sorts things in the environment, orders them, and makes judgments about attributes, he discovers relationships that are more sophisticated. For example, the basic concepts of simple succession and then sequence are developed to a higher level, enabling the student to form complicated series.

When the child makes a decision about members of a class or of a set, he makes two kinds of judgments. One will lead to *numerical* ideas—for example, whether two or more sets are equivalent; which set has more or fewer members; which transitivity relationships exist among the sets. The other type of judgment is *nonnumerical*. Do objects belong together because of similarity of some attribute(s), or are the objects different and thus do they form disjoint sets or classes? Are there sets with no objects?

A *set* can be defined by listing examples of its members (exemplars). This is definition by *extension*. Definition by *intension* "specifies the properties shared by members of the class" (Gruber & Voneche, 1977, p. 359). As cognition develops, individuals extend their ideas regarding sets. They come to realize that members of a set need not have an inherent common attribute, but need only to be separated in thought from other elements. This idea was expressed in Chapter 7, in which concepts were constructed by first comparing equivalent sets comprised of items with the same attributes. Finally, common perceptible attributes were eliminated, and all that remained was the equivalence relation. Thus, the idea of *set* progresses from sets whose elements are things to abstract, unlimited sets, such as the set of *multiples of five* or the set of *counting numbers* or the set of *simple closed curves*. Sets can be viewed as hierarchical; the members of a set are, themselves, sets.

Subset and *class inclusion* relationships may be representd in a concrete manner, with diagrams, or by symbols. The mode of representation will modify the difficulty level of the task. In Figure 8.2, a boundary is drawn around members of a set. A concrete representation would stand these children inside boundaries (of yarn or chalk, for example). The more abstract manner of diagramming set relationships employs Venn diagrams, as in Figure 8.3, which shows the relationship of the set of girls to a class of boys and girls. (The brackets show that we have a *set of children* in mind, rather than individual boys and girls.)

Children with glasses
Children with curly hair
Children who are girls

Figure 8.2

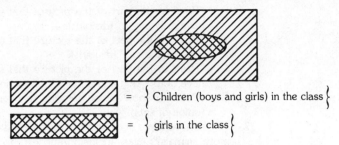

= { Children (boys and girls) in the class }

= { girls in the class }

Figure 8.3

The Assessment and Teaching Activities that follow are designed to help the teacher ascertain answers to the following questions:

What mathematics *can* this child learn?
What mathematics that he can learn *should* he put most energy toward?

*Assessment
Activities:*

1. Say, "Raise your hand if you have a dog." Observe if child responds to this unambiguous direction.

2. Say, "Raise your hand if you have a dog, and keep it down if you do not have a dog" (disjoint sets). Observe if child becomes confused in this direction.

3. Say, "Raise your hand if you have a dog or a cat" (union of sets).

4. Say, "Raise your hand if you have a dog and a cat" (intersection of sets).

5. Say, "Raise your hand if you have a dog, and keep it down if you have a cat." Observe what the child with both a dog and cat (as in item 4) does. This child will be confused if he recognizes that he falls into the intersection of those with dogs and cats. Observe how he solves this dilemma. Does frustration occur? Does the child focus on either dog or cat? Is the child aware of conflict?

6. Given the following diagram (Figure 8.4), ask the child to explain the set relationships as they respond to the following questions:

Figure 8.4

    A. "Point to the part of the picture that stands for families who have only a dog" (disjoint).
    B. "Point to the part of the picture that stands for families who have only a cat" (disjoint).
    C. "Point to the part of the picture that stands for families who have both a dog and a cat" (intersection of sets).
    D. "Point to show which families have a dog, a cat, or both" (union of sets).
  7. "If you have three dogs and two cats, would you have more dogs or more animals?" (class inclusion/subset).
  8. "If you have three dogs, two cats, and four marbles, would you have more marbles or animals?" (class inclusion/subset).

Teaching Activities

Age appropriateness for both the Assessment and Teaching Activities is determined by functional limitations based upon the student's inability to meet age-specific expectations. Thus, an activity that may be appropriate for a 7-year-old who is very bright may be a challenge for an older child who is developing at a slower cognitive rate. This is in keeping with the functional model, which focuses on characteristics of a disability or talent and their impact on a student's ability to learn and to function.

> *Instructional Strategy 8.2.1:*
> *Control for component skills.*

The ability to enumerate may inadvertently *interfere* with a child's ability to show class inclusion and may give the teacher an inaccurate view of the child's performance on class-inclusion tasks. For example, when the child is asked, "Which is more, the flowers or the roses?" the child may compare the number of items in these two classes by enumerating. Counting involves tagging each object in a one-to-one matching (Wilkinson, 1976). Once an item has been counted, it is removed from the "to-be-counted" category. But to do so (that is, to remove a counted item) is contrary to what the child would have to do in a class-inclusion task. Roses, having already been counted, cannot be counted again without violating a principle of enumeration: Count each element *once and only once.* What is the child to do? He counts what is left over—he counts the other subset. Present the following sequence (Conditions A–E) to control for interference of the child's ability to count.

Condition A: Present three roses, two tulips, one carnation. Ask, "Which is more, the flowers or the roses?"

      Ⓡ   Ⓡ   Ⓡ   Ⓣ   Ⓣ   Ⓒ

Condition B: Present three roses, one tulip, one carnation. Ask, "Which is more, the flowers or the roses?"

$$\text{(R)} \quad \text{(R)} \quad \text{(R)} \quad \text{(T)} \quad \text{(C)}$$

Condition C: Present three roses, three tulips. Ask the same question.

$$\text{(R)} \quad \text{(R)} \quad \text{(R)} \quad \text{(T)} \quad \text{(T)} \quad \text{(T)}$$

Condition D: Present three roses, two tulips, two carnations. Ask the same question.

$$\text{(R)} \quad \text{(R)} \quad \text{(R)} \quad \text{(T)} \quad \text{(T)} \quad \text{(C)} \quad \text{(C)}$$

Condition E: Present six roses, two tulips. Ask the same question.

$$\text{(R)} \quad \text{(R)} \quad \text{(R)} \quad \text{(R)} \quad \text{(R)} \quad \text{(R)} \quad \text{(T)} \quad \text{(T)}$$

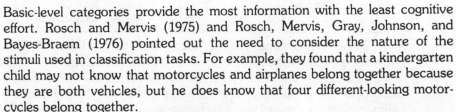

> **Instructional Strategy 8.2.2:**
> *Use natural categories (basic level) for classification tasks rather than arbitrary categories.*

Basic-level categories provide the most information with the least cognitive effort. Rosch and Mervis (1975) and Rosch, Mervis, Gray, Johnson, and Bayes-Braem (1976) pointed out the need to consider the nature of the stimuli used in classification tasks. For example, they found that a kindergarten child may not know that motorcycles and airplanes belong together because they are both vehicles, but he does know that four different-looking motorcycles belong together.

The child who is slow should be allowed to categorize basic-level objects into basic categories, rather than into superordinate categories. Examples of superordinate categories and the related basic-level categories are shown in Table 8.2.

Table 8.2. *Classification Structures*

Superordinate	Basic Level
Clothing	Shoes, socks, shirts, pants
Furniture	Tables, chairs, beds, dressers
People's faces	Men, women, young girls, infants
Vehicles	Cars, trains, motorcycles, airplanes

A classification task on basic-level objects would involve presenting the child with four assorted shirts, four different tables, four human faces, and four different cars (or photographs of objects). The child would be told to form groups of objects that belong together.

A superordinate classification involves giving *one* picture of *each* of the four objects in each of the four superordinate categories—one car, one train, one motorcycle, one airplane; one table, one chair, and so on. In this task, a child must put objects together to form superordinate groups. This calls for a higher level of abstraction than the reasoning involved in abstracting basic-level concepts.

> ### Instructional Strategy 8.2.3:
> *Distinguish between intensional and extensional aspects of concepts.*

The *intensional* aspect of a concept provides the criteria that form the category or class. The *extensional* aspect refers to the elements that satisfy the defining criteria (that is, examples of the class). Table 8.3 shows a comparison of the intensional-extensional dichotomy:

Table 8.3. *Intensional–Extensional Classification*

Intensional	Extensional
Birds	Robin, chicken, sparrow
Flowers	Rose, tulip, carnation
Vehicles	Bicycle, bus, car
Fruit	Grape, banana, peach

This instructional strategy (suggested by Markman & Siebert, 1976) involves essentially the same type of classification task as in IS 8.2.2. Here, the idea of a category and its examples are emphasized, with a focus on placing examples into a category. The following activity includes this idea but goes a step further—a more basic classification activity is involved. This includes grouping examples of an example, and is exclusively extensional. This idea is shown in Figure 8.5, as an expansion from Table 8.3.

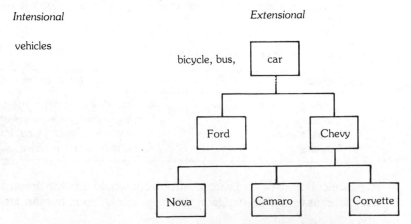

Figure 8.5. *Expanded Extensional Classification*

Thus, within the extensional category, a child is asked to give examples of a car and then to give examples of one of the examples. This part of the task has a deductive nature; the child proceeds from the general to the specific.

> **Instructional Strategy 8.2.4:**
> *Distinguish between classes and collections.*

Markman and Siebert (1976) distinguished between *classes* and *collections* and the types of concepts they characterize. In the previous two instructional strategies, we were dealing with *classes* at different levels of inclusion (for example, superordinate or intensional: fruit; extensional: apple; basic level: Red Delicious, Yellow Delicious, MacIntosh, Cortland). Each object could be considered independently, and its class membership could be determined by considering it in terms of its defining properties. But to determine whether an object is a member of a collection, its relationship to other objects must be considered. A *bushel* of apples is a bushel of apples only if there is close spatial proximity between an agreed-upon amount of apples; likewise a *bunch* of bananas. An object, by itself, cannot be judged a member of a collection, but it can be judged a member of a class. The Yellow Delicious is a member of the class *apple*, whether or not it belongs to the collection *bushel*. Markman and Siebert suggested that objects and collections are similar in that each relies upon relationships among parts for identity. For example, if a human body is to be considered as an object—as a human body per se—it must have a specific organization of its parts.

These ideas suggest a modified approach to the Piagetian model for class inclusion that involves the following questions:

Piaget Condition: Given red and blue blocks arranged in a pile, the child is asked: Would the child who owned the blue blocks or the child who owned the blocks have more blocks to play with?

Modified Condition A: Given the same blocks, the child is asked: Would the child who owned the blue blocks or the child who owned the *pile* of blocks have more blocks to play with?

Modified Condition B: Given six roses and two tulips, the child is asked: Would a woman who was given the roses or a woman who was given the *bunch* of flowers receive more flowers?

Modified Condition C: Given three cups and two saucers, the child is asked: Are there more cups or more dishes in the *set* of things to eat with?

Markman and Siebert found that children who failed the Piaget condition were successful in the modified conditions.

> ### Instructional Strategy 8.2.5:
> ### Use good exemplars of a class.

Some examples of a class are better exemplars than others. Horses and dogs have been found to be better examples of the animal class than flies and bees. Children do better when asked the class-inclusion question using horses and dogs (Condition A) than when using horses and flies (Condition B) or dogs and bees (Condition C), as in the following questions:

Condition A: Given pictures of horses and dogs, ask: Are there more horses or animals?

Condition B: Given pictures of horses and flies, ask: Are there more horses or animals?

Condition C: Given pictures of dogs and bees, ask: Are there more dogs or animals?

This instructional strategy also facilitates development of the negative idea:

Condition A: Given pictures of horses and dogs, say: Hand me all of the pictures that do *not* show dogs.

Condition B: Given pictures of horses, dogs, and bees, say: Hand me pictures of things that do *not* fly.

> ### Instructional Strategy 8.2.6:
> ### Restrict the number of exemplars in classification tasks.

Presenting fewer examples of a category was found to enhance the use of commonality of category membership (Worden, 1976). Worden (1975) also suggested that students were able to classify because they were allowed to do the organizing, rather than being required to classify according to abstract, formal rules organized by the adult. Thus, if you want a slower child to abstract the idea of cubes as a subset of blocks, restrict the number of geometric shapes in the complementary set.

Condition A: Present six cubes and four tetrahedrons rather than six cubes, one tetrahedron, two spheres, and one rectangular shape. Ask: Are there more cubes or blocks?

Condition B: Present four cubes and six tetrahedrons. Ask: Are there more cubes or blocks?

116

8.3.0 Number Operations and Relationships

The operations and relations that hold between sets serve as a model for number operations and relations.

The set operation of union of two disjoint sets is another way of stating the *logical addition of classes*. If two different classes are combined, a new class results; for example, the class of dogs combined with the class of cats results in the class of animals (dogs and cats). This, in turn, is related to the arithmetical operation of adding numbers. However, logical addition of classes is a more general idea, since no counting is involved. The objects in the sets are actually combined, rather than combining the number of each set.

The intersection of sets is another name for *logical product* which results from *logical multiplication of classes*. For example, as a child selects the defining property of a set, he or she then collects objects belonging to that set. Soon, this is done in more than one way, and the child considers different attributes while at the same time forming the class of things which belong to both of two classes.

An example of logical multiplication of classes is considering a class of white marbles or a class of glass marbles. The logical multiplication of these sets yields the class of white glass marbles. This may be extended by considering additional classes, such as yellow marbles, black marbles, and plastic marbles. This gives rise to six classes resulting from three colors and two textures (see Table 8.4).

Table 8.4. *Six Classes of Marbles*

Color	Texture	
	Plastic	Glass
White	X	X
Black	X	X
Yellow	X	X

This is a model for the cross-product idea of multiplication (see Appendix A for other multiplication models). Understanding logical multiplication of relations underlies the transformation-invariant relationships embedded in the conservation tasks discussed earlier in this chapter; for example, both the length and width of the transformed ball of clay must be considered simultaneously in order to realize that the mass remains the same.

Assessment Activities:

The following activities assess number operations and number relations. See Appendix A for a review of *factor* and *equivalence*.

1. Complete the following by writing the operation sign that makes each statement true:
   a. $3 \square 5 = 8$
   b. $4 \square 3 = 12$
   c. $14 \square 2 = 7$
   d. $9 \square 2 = 7$
   e. $(3 + 5) \square (1 + 3) = (14 - 2) + 0$

2. Mark X by the equivalence relations.
    a. goes to the same school as   (x)
    b. is taller than
    c. is the same age as   (x)
    d. is the same shape as   (x)

3. Show the following relation by completing the diagram:

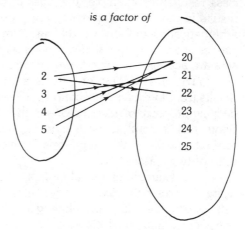

4. Complete the graph by marking x's to show the following relation:

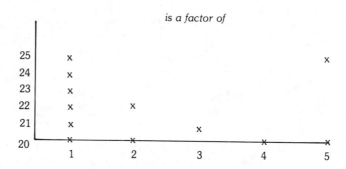

Teaching Activities

The instructional strategies that follow should in no way be thought of as a total curriculum for number operations and relations. These strategies represent some modifications for students with special learning needs. The teacher should try these instructional strategies with related topics and with a variety of learners.

Instructional Strategy 8.3.1:
Provide immediate knowledge of results.

Inform the slower-learning student of the correctness of his response through consistent and immediate feedback. In addition to the use of incentives (food, toy, token), employ spoken words (*no, wrong, good, that's right*) or a signal (light or bell) for a correct answer, and a buzzer for an incorrect answer.

Ask: How do you find the number of balls when you have two balls and one more?

Possible answers: I count. (*Good, how else?*)
I take away. (*No, that's not right. What happens when you take away?*)
I put together. (*Yes, that's good, and how do you find the number?*)
I add. (*Good, and how many do you have?*)
Three. (*Very good.*)

> ## Instructional Strategy 8.3.2:
> ### Use distributed repetition.

Denny (1966) suggested that there is more gain in overlearning for retarded individuals than for normals. "This gain is greatest when the instruction is not massed or concentrated but spaced within teaching periods and across periods" (Mercer & Snell, 1977, p. 148). Positive transfer is facilitated because the chances for using the particular idea—such as number operation or number relation (equivalence)—in more than one setting is facilitated. Thus, it appears that identifying a number operation such as addition in everyday settings will provide distributed practice in the mathematical relationship.

For the gifted child, the teacher might ask for environmental examples of addition or of other operations, as well as of equivalence relations. These settings may then be used for distributed context learning for slower learners. The gifted child thus becomes a model for identifying number ideas in context settings.

A tape recorder can be used for practice with slower students. The teacher could present an addition situation and have the student select a picture card that it represents. For example, the tape says: "Put the picture in the box [or any specified place] that shows two boys who have the same number of sisters" (see Figure 8.6). Picture C should be selected.

> ## Instructional Strategy 8.3.3:
> ### Use verbal rehearsal.

Verbal rehearsal involves labeling. The child should be encouraged to label number operations as he performs them: "I am adding," "These sets have the same number." Rehearsal strategies help to circumvent short-term memory deficiency (Ellis, 1970).

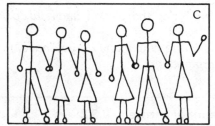

Figure 8.6

(IS)
> **Instructional Strategy 8.3.4:**
> *Facilitate attention to the relevant dimension.*

Teaching strategies summarized by Mercer and Snell (1977, p. 315) that relate to selecting the appropriate operation include the following:

> *Reduce the number of irrelevant dimensions.*
> *Increase the number of relevant dimensions.*
> *Reinforce attention to relevant dimensions.*
> *Make relevant cues distinctive.*

(IS)
> **Instructional Strategy 8.3.5:**
> *Plan for transfer in learning.*

Transfer means that knowledge acquired in one context is used in a related setting—for example, a student may use the one-to-one relation and the addition operation to help him select equivalence relations. Identifying that the relation *is the same number as* may be represented as shown in Figure 8.7.

(IS)
> **Instructional Strategy 8.3.6:**
> *Employ Torrance's Hierarchy of Creative Skills*

Torrance (1970) utilized the following activities in the Ginn Reading 360 Program. These skills are applied to mathematics in this IS.

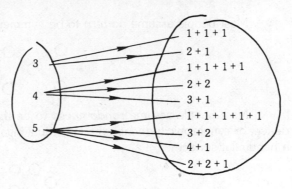

Figure 8.7

Level 1 Instructional Application
Produce new combinations through manipulation. Utilize the number operations to analyze number by producing many combinations under the relation *is the same as*:

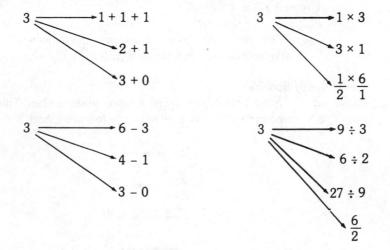

Level 2 Application
See and produce increasingly larger numbers of possibilities in combining objects. Arrange five coins in a linear (orderly) array, a patterned array, and an accidental or random array. Most people assume *order* to be sequential, linear and functional:

○
○
○
○
○

121

Most people assume *pattern* to be symmetrical, lateral and decorative:

Most people assume *randomness* to be the result of accident, coinci-dence, or synchronicity (hence the individual invariably throws the coins onto a handy flat surface):

Level 3 Application
Order sequences of events. This activity may be applied to investigating the effect of number operations. This activity is an organizer for studying the associative property.

Given $3 + 3 \times 3 - 3 \div 3$

Ask: "What are all the possible answers to this sequence?"
(Possible answers include: 0, 5, 17.)

Level 4 Application
Make easy, simple predictions from limited information. This activity may combine a transformation with a relation, as follows ("add 7"):

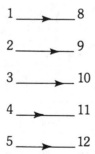

Level 5 Application
Develop ability to produce alternate causes of behavior.

Ask: "What will happen if the number 5 is squared four times?"
Ask: "What will happen if a number is consecutively doubled three times and halved three times?"

Level 6 Application
Go beyond the obvious. Torrance (1970, p. 50) described children at this stage of development (see Endnote 1) as being able to get a "maximum of information from a small number of clues." In Figure 8.8, the child is to complete the pattern and state the relation (square).

122

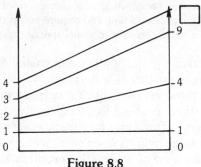

Figure 8.8

8.4.0 Mathematics as a Language of Relationships

Many relationships involve quantitative ideas. In addition to the obvious use of mathematical symbols to express mathematics relationships—for example, $=$, $-$, $>$, $<$, $\leq$ —expressed by inflectional endings.

Examples

Comparative:	longer, shorter, heavier
Superlative:	longest, shortest, heaviest
Singular:	apple, table, house
Plural:	apples, tables, houses

Other mathematical relationships expressed in general language include the topological ideas of order, proximity, enclosure, and separateness.

Examples

Order:	My *first* child.
	This is my *third*.
	The plane is landing *first*.
Proximity:	That was a *near* miss.
	He is *far away*.
Enclosure:	The doll is *in the box*.
	He is standing *inside*.
	The house *between* the buildings burned.
Separateness:	The boat was cut *apart* from its moorings.

Many prepositions express mathematical relationships—on, under, above, below, up, down, behind, and so forth.

Language that expresses mathematical relationships is particularly troublesome for learning-disabled children (Wiig & Semel, 1976):

> The comprehension of prepositions and prepositional phrases appears particularly sensitive to deficits in spatial perception and in the coding of visual-spatial events. . . . [Locative] prepositions denoting a definite, static position or state, with respect to location, quality, condition, or possession appear easiest for the learning disabled child to interpret. . . . Prepositions noting a change in direction or condition (directional prepositions) are generally harder to code linguistically

and interpret semantically for the learning disabled child. They imply movement or change in space and time and require fine distinctions of space and time which may render them sensitive to visual-spatial deficits. (p. 50)

In addition to quantitative and spatial relationships, we may note a third aspect of mathematics as a language of relationships: temporal relationships, particularly those expressed by tense (past, present, future). Let us take, as an example, a word problem that involves the additive idea of subtraction. In the word problem, "Three plus how many equals five?" the mathematics sentence translates into an indirect open sentence: $3 + \square = 5$. The future tense is implied in this additive type of subtraction problem, although *equations are read in the present tense*. We do not say, "three plus two was five" or "three plus two will be five"; rather, we say, "three plus two is five." Word problems that involve past, future, or mixed tenses (see Table 8.5) clearly would present problems for children with deficits in language competence who have to translate the problems to present-tense equations. This is especially true because most children are unaware of the discrepancy between word-problem tenses and the present tense of equations.

Table 8.5.   *Typical Word Problems in Elementary Mathematics Texts*

Tense	Word Problem
Past	Johnny had 3 apples. He ate 1. How many were left?
Present	Johnny has 3 apples. He eats 1. How many are left?
Future	Johnny is going to buy 3 apples. He will eat 1. How many will he have?
Mixed	Johnny had 3 apples. If he eats 1 how many will he have left?
Past-Comparative	Johnny had 3 apples and Martha had 7. How many more did Martha have?
Present-Comparative	Johnny has 3 apples and Martha has 7. How many more does Martha have?
Future-Comparative	Johnny is going to buy 3 apples and Martha is going to buy 7. How many more will Martha have?
Present-Conditional	Johnny has 3 apples. If he eats 1, how many does he have?
Mixed-Conditional	If Johnny had 3 apples and he eats 1, how many will he have left?

Assessment Activities: Piaget suggested that syntactic and semantic aspects of language play a part in accelerating the processes of both classification and seriation (Gruber & Voneche, 1977). Therefore, assessment of a child's performance in dealing with mathematics as a language of relationships is helpful in assessing competence in both mathematics and language.

124

1. Locative and directional prepositions and spatial relationships:
   A. Circle the child who is on top of the hill.

   B. Circle the child who is going up the hill.

   C. Circle the child who is in the middle of the line to climb the slide.

2. Topological relationships:
   A. *Proximity*—Mark the ball close to the boy.

B. *Separateness*—Mark the tree trunk that is separate from its leaves.

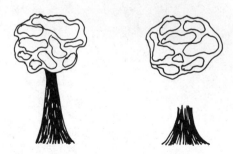

C. *Order*—Mark the child who is first in line to climb the ladder.

D. *Enclosure*—Mark the plane that is inside the circle.

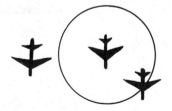

3. Quantitative language:
    A. *Comparative*—Mark the shorter boy.

B. *Superlative*—Mark the happiest face.

C. *Plural-Singular Recognition*—Mark the mathematical sentences that describe the picture.
Mark the mathematical sentence that describes the picture.

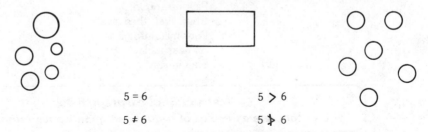

5 = 6   5 > 6

5 ≠ 6   5 ≯ 6

If the student is not aware of the difference in inflectional endings of these two questions (3c), she may answer these items incorrectly. The first example requires both 5 ≠ 6 and 5 ≯ 6, while the second example asks for only one response (5 ≠ 6 or 5 ≯ 6). Plurals and singulars may be a confusing element in languages whose plurals are not necessarily designated by the letter s. For example, in the sentence "I have three sister" the idea of plurality is logically expressed by the word "three." To say "three sisters" is a redundancy that does not follow the logic of some languages.

4. Tense:
Present word problems modeled after those included in Table 8.5. Note which types the student can do correctly. Relate the content of the problems to the experiences and interests of the student; use names of his friends, situations that are real. Use this data to select appropriate instructional strategies that involve word problems.

Teaching Activities

Learning disabilities are often described as language-based disorders. Therefore, some students will need circumvention of their language problems in order to facilitate learning of specific quantitative, spatial, or temporal relationships that are typically expressed in general language. Other students will not be able to sort out complicated syntactic forms, and those students at a slower rate of intellectual development will need basic, concrete representations and models as aids. The instructional strategies that follow include suggestions for selection of curriculum as well as pedagogical modifications.

Instructional Strategy 8.4.1:
Use shorter, simpler sentences when giving directions.

Some children need help in attending to mathematical relationships expressed in general language. Use of shorter, simpler directions facilitates taking note of these relationships.

Examples

> Put on coat.
> Pick up spoon.
> Chew first, then swallow.
> Pour milk into glass.
> Sit near table.
> Stay in yard.

(IS)

Instructional Strategy 8.4.2:
Use concrete examples of spatial and quantitative relationships.

For teaching prepositions that represent mathematical relationships, accompany language instruction with concrete examples. If the spatial idea of *between* is to be learned, direct the child to place an object between two other objects, as in Figure 8.9.

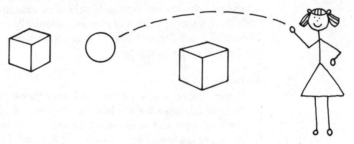

Figure 8.9.   *Child Places Ball Between Blocks*

(IS)

Instructional Strategy 8.4.3:
Use prompting.

The teacher may prompt a child's response for the situation, "Put the doll in the box." The teacher may need to correct the child who puts the doll on top of a box turned upside down. The teacher places child's hands holding doll *into* the open box.

(IS)

Instructional Strategy 8.4.4:
Use relational rebus symbols.

Relational rebus symbols were presented in Chapter 1, page 8, and included the following:

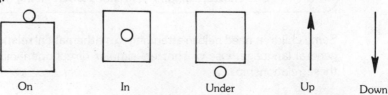

| On | In | Under | Up | Down |

128

Instructional Strategy 8.4.5:
Incorporate modeling plus verbal instructions.

Mercer and Snell (1977, p. 253) summarized research that supported this instructional strategy: "Altman and colleagues (1972) found that modeling plus verbal instructions are needed to provide the SMR [severe mentally retarded] observer with as many cues (verbal and motor) as possible when demonstrating a task [for example, directional words—up, down]."

Instructional Strategy 8.4.6:
Highlight inflectional endings.

The use of color, different-size print, underlining, or clustering are techniques that cue the child to attend to inflectional endings that change mathematical meaning.

Examples
    short**er**
    short**er**
    short*er*
    short er

Instructional Strategy 8.4.7:
Introduce novelty into learning experience.

This activity is particularly appropriate for the creative student, since the strategy allows for new and original relationships. Present the four-by-four square as shown in Figure 8.10. The 16 words can generate 256 sentences,

YOUNG	MAN	LEARNS	QUICKLY
ROUND	BALL	BOUNCES	SIDEWAYS
STONE	BUILDING	SEEN	STANDING
BLUE	STAR	TWINKLES	NIGHTLY

Figure 8.10. *Pattern of Thinking Square*

From Keith Albarn and Jenny Mial, *Diagram, The instrument of thought.* London: Thames and Hudson, Ltd., 1977. Used with permission.

each containing four words. The sentences will generate images that may be judged according to three criteria: sensibility, interest, and acceptability. The child can then graph his sentence patterns.

The sentence generated in Figure 8.11, "Blue man seen nightly," is shown in Pattern B of Figure 8.12. This sentence generates an image in one's mind that satisfies all three criteria: The sentence is sensible, interesting, and acceptable. The choice of route from word to word reveals the pattern of thinking. Orderly, linear patterns show attention to symmetrical forms that the individual is most familiar with and have the most meaning in terms of experience. Asymmetrical routes show lateral thinking, exploration of new possibilities. These are routes based on either a visual or a verbal pattern.

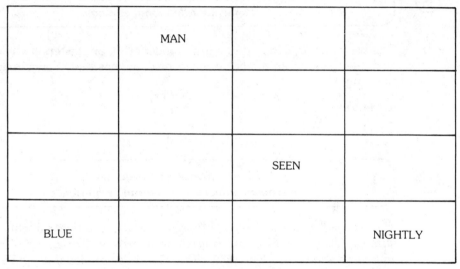

	MAN		
		SEEN	
BLUE			NIGHTLY

Figure 8.11.   *Example of Sentence that Generates an Image that Is Sensible, Interesting, and Acceptable*

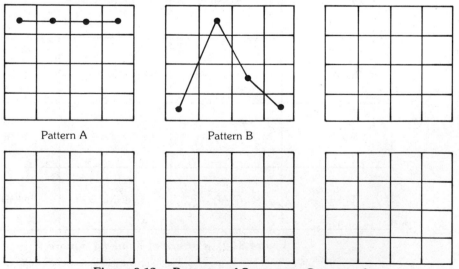

Pattern A   Pattern B

Figure 8.12.   *Patterns of Sentences Generated*

Instructional Strategy 8.4.8:
Show how different verbal directions may change spatial relationships.

Arrangement of objects according to various criteria provides a concrete activity for forming relationships. Give a child five pennies and ask him to make arrangements in space. (Note that this activity is the same as for IS 8.3.6, but the purposes of these two instructional strategies are different.)

Orderly Array
"Put the pennies in an orderly arrangement."

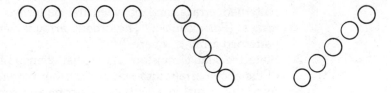

Patterned Array
"Make a pattern with the pennies."

Random Array
"Throw the pennies on the table."

Summary Thought Questions and Activities

I. Reveal your pattern of thinking for the following by generating and graphing sentences:

SMALL	CHILD	DEVELOPS	FAST
SQUARE	BLOCK	BURNS	NICELY
GLASS	WINDOW	FALLS	UPWARDS
YELLOW	DIAMOND	SPARKLES	BRIGHTLY

RETARDED	LADY	WRITES	LIGHTLY
BRICK	HOUSE	OBSERVED	BURNING
GIFTED	BOY	SOLVES	SIDEWAYS
CREATIVE	FISH	SWIMS	DAILY

131

  A. State which of the criteria you used for generating sentences (sensible, interesting, acceptable).
  B. Describe the image each sentence created.
  C. Categorize your patterns: (a) orderly, linear, symmetrical; or (b) asymmetrical. Were the patterns consistent *within* a grid? Was there consistency across grids? Is there any relationship between a dominant grid pattern and the way you form relationships?
II. Arrange a set of five objects to form the following arrays:
  A. Form as many *orderly* arrangements as you can.
  B. Form as many *patterned* arrays as you can.
  C. Form an *accidental* array.
    1. Describe the spatial relationships between the orderly and patterned arrays, and between the patterned and accidental arrays. (For example, "The orderly array is vertical, while the patterned array is square.")
    2. State the spatial transformations that changed the form of one of the above arrays into one of the other patterns.
    3. State the spatial invariants that accompany some or all of the transformations; for example, "The number [size, color, . . .] remained the same."
III. In ten minutes write your answers to the following questions taken from Luria (1973). These have been categorized according to the mathematical relationships expressed by the language.
  A. Comparative-Superlative
    1. Olga is lighter than Kate but darker than Sonia. Who is lightest?
    2. John is taller than Peter. Which boy is shorter?
    3. Mark the square that is less dark.

Pattern A   Pattern B

  B. Sequencing (involving word meaning, cause–effect, spatial relationships, prepositions)
    1. I had breakfast after I had sawed the wood. What did I do first?
    2. Nicky was struck by Peter. Who was the bully?
    3. Peter struck John. Who was the victim?
    4. Underline the correct statement:

        the sun is lit by the earth   the earth is lit by the sun

    5. Circle what is used to start a fire:

                  stove   firewood   matches

    6. Underline the correct phrase:

spring comes before summer   summer comes before spring

7. Draw the following series—two circles, a cross, and a dot.
8. Draw a dot beneath a triangle.
9. Draw a cross on the right of a circle but on the left of a triangle.
10. Draw a circle beneath a cross.
11. Draw a circle to the left of a cross.
12. Draw a cross beneath a circle.
13. Draw a triangle below a circle.
14. Draw a triangle to the right of a cross.

C. Word Meaning (opposite, double negative)
  1. If it is now night, place a cross in the white square, and if it is now day, place a cross in the black square.

  2. I am unaccustomed to disobeying rules. Is this the remark of a disciplined or an undisciplined person?

D. Possessives
  1. How would "your father's brother" be related to you?
  2. How would "your brother's father" be related to you?
  3. Is "sister's mother" equal to "grandmother"?

    yes   no

  4. Is "father's brother" equal to "brother's father"?

    yes   no

  5. Does "the dog's master" equal "the master's dog"?

    yes   no

  6. You are given pictures of a woman and a girl. Circle the "mother's daughter."

E. Assessing Dominance Relationships
1. Interlock your fingers by clasping your hands. Which thumb is closest to your body?

left hand thumb   right hand thumb

2. Which hand do you throw a ball with?

left hand   right hand

3. Compare the fingernails of your little fingers. Which hand has the wider little fingernail?

left hand   right hand

4. Which eye do you use to look through a tube?

left eye   right eye

5. Which leg plays the most active role in kneeling?

left leg   right leg

6. Cross your arms in front of you placing them on your chest. Which arm is on top?

left arm   right arm

7. Which hand is mainly used when you are working?

left hand   right hand

8. Which hand plays the active role when you clap?

left hand   right hand

9. Who in your family is (was) left-handed? List blood relations.

F. Spatial Relationships (field dependent-independent; reflections)
1. Trace the shape of Figure A in Figure B.

Figure A   Figure B

2. Draw the mirror image of the following:

G. Reverse Order
1. Count backward from 100 in 7's.
2. Count backward from 100 in 13's.

From *The Working Brain: An Introduction to Neuropsychology*, by A. R. Luria, pp. 336–340, © Penguin Books Ltd. 1973, translation © Penguin Books Ltd. 1973, Basic Books, Inc., Publishers, New York.

IV. Use your responses to the Luria tasks in activity III above as diagnostic data. Answer the following questions:
A. Did you feel rushed because of the time limit?
B. Did you draw diagrams to help you figure out some of the relationships?
C. Did you have to reread some of the questions more than others? Was there a pattern to these?
D. Did you do better on particular categories of questions? Was there a pattern?
E. Which questions were most troublesome because of the time limit?
F. Were your dominance relationships consistently either left or right, or was there mixed dominance?
G. Were you aware of differences in your performance between activity G1 and activity G2, involving counting backwards?

Suggested Readings

Arieti, S. *Creativity, the magic synthesis.* New York: Basic Books, 1976.

Baer, D. M., & Guess, D. Teaching productive noun suffixes to severely retarded children. *American Journal of Mental Deficiency*, 1973, 77, 498–505.

Bartlett, E. J. Sizing things up: The acquisition of the meaning of dimensional adjectives. *Journal of Child Language*, 1976, 3, 205–219.

Brewer, W. A., & Stone, J. B. Acquisition of spatial pairs. *Journal of Experimental Child Psychology*, 1975, 19, 299–307.

Clark, H. H. The primitive nature of children's relational concepts. In J. R. Hayes (Ed.), *Cognition and the development of language.* New York: John Wiley & Sons, Inc., 1970.

Clark, H. H. Space, time, semantics and the child. In T. E. Moore (Ed.), *Cognitive development and the acquisition of language.* New York: Academic Press, 1973.

Cohn, R. Arithmetic and learning disabilities. In H. R. Mykelbust (Ed.), *Progress in learning disabilities, volume II.* New York: Grune and Stratton, 1971.

Donaldson, M., & Balfour, G. Less is more: A study of language comprehension in children. *British Journal of Psychology*, 1968, 59, 461–471.

Farnham-Diggory, S. *Learning disabilities: A psychological perspective.* Cambridge, Mass.: Harvard University Press, 1978.

Ginsburg, H. *Children's arithmetic: The learning process.* New York: Van Nostrand, 1977.

Miller, G. A., & Johnson-Laird, P. N. *Language and perception.* Cambridge, Mass.: Harvard University Press, 1976.

Novillis, C. F. *Children's meanings of some relational and dimensional terms.* Paper presented at the Sixth National Conference on Diagnostic and Prescriptive Mathematics, Tampa, April 23, 1979.

Torrance, E. P. *The search for satori and creativity.* Buffalo: Creative Education Foundation, Inc., 1979.

Torrance, E. P., & Meyers, R. E. *Creative learning and teaching.* New York: Dodd, Mead and Company, 1970.

Wallas, G. *The art of thought.* New York: Harcourt, Brace, 1926.

References

Cohen, J. Subjective time. In J. T. Fraser (Ed.), *The voices of time.* New York: George Braziller, 1966.

Denny, M. R. A theoretical analysis and its application to training the mentally retarded. In N. R. Ellis (Ed.), *International review of research in mental retardation* (Vol. 2). New York: Academic Press, 1966.

Ellis, N. R. Memory processes in retardates and normals. In N. R. Ellis (Ed.), *International review of research in mental retardation* (Vol. 4). New York: Academic Press, 1970.

Gruber, H. E., & Voneche, J. J. *The essential Piaget.* New York: Basic Books, 1977.

Keyser, C. J. *Mathematical philosophy.* New York: E. P. Dutton & Company, 1922.

Markman, E. M., & Siebert, J. 1976. Classes and collections: Internal organization and resulting holistic properties. *Cognitive Psychology*, 1976, *8*, 561–577.

Mercer, C. D., & Snell, M. E. *Learning theory research in mental retardation: Implications for teaching.* Columbus, Ohio: Charles E. Merrill Publishing Company, 1977.

Piaget, J. *The child's conception of number.* New York: W. W. Norton, 1965.

Piaget, J. *On the development of memory and identity, volume II, 1967 Heinz Werner lecture series.* Barre, Mass.: Clark University Press, 1968.

Reisman F. K. *An evaluative study of cognitive acceleration in the early years.* Unpublished doctoral dissertation, Syracuse University, 1969.

Reisman, F. K. *A guide to the diagnostic teaching of arithmetic* (2nd ed.). Columbus, Ohio: Charles E. Merrill Publishing Company, 1978.

Reisman, F. K., & Torrance, E. P. *Comparison of children's performance on the Torrance test of creative thinking and selected Piagetian tasks.* Paper presented at the First World Congress on Future Special Education, Stirling, Scotland, June 25-July 1, 1978.

Rosch, E., & Mervis, C. B. Family resemblances: Studies in the internal structure of categories. *Cognitive Psychology*, 1975, *7*, 573–605.

Rosch, E., Mervis, C. B., Gray, W. D., Johnson, D. M., & Bayes-Braem, P. Basic objects in natural categories. *Cognitive Psychology*, 1976, *8*, 382–439.

Torrance, E. P. *Encouraging creativity in the classroom.* Dubuque, Iowa: William C. Brown Company, 1970.

Wilkinson, A. Counting strategies and semantic analysis as applied to class inclusion. *Cognitive Psychology*, 1976, *8*, 64–85.

Worden, P. E. Effects of sorting on subsequent recall of unrelated items: A developmental study. *Child Development*, 1975, *46*, 687–695.

Worden, P. E. The effects of classification structure on organized free recall in children. *Journal of Experimental Child Psychology*, 1976, *22*, 519–529.

Endnote

1. Reisman and Torrance (1978) found significant relationships of children's performance on selected Piagetian tasks of conservation and on the Torrance Tests of Creative Thinking. Ability to engage in imagery, flexibility of thought, and resistance to premature closure, in particular, appear to underlie both creative thinking and conserving. The results of this study suggest additional theoretical explanations for conservation in relationship to research in creative thinking.

Instructional Strategies Based on Cognitive Needs: Higher-Level Generalizations

9

When one considers the usual definition of a generalization—putting two or more concepts together into some relationship—it becomes apparent that the topics included in this chapter are much more complex and abstract than usually considered. Yet many of these topics are typical of the kindergarten and first grade mathematics curriculum; many are trouble spots.

9.1.0 Basic Facts

Basic facts are higher-level generalizations, because they are comprised of concepts (numbers) and a relationship (addition, multiplication, subtraction, division). Basic facts result from performing a binary operation on two numbers that are represented by a single digit, such as

$$3 + 5 = 8 \text{ or } 7 \times 6 = 42.$$
The inverses of these are also considered basic facts:

$$8 - 3 = 5 \text{ or } 42 \div 7 = 6.$$

Assessment Activities:

1. Have child complete grids such as those shown in Figure 9.1.

(a)

+	0	1	2	3
0	0	1		
1				
2			4	
3			5	6

(b)

x	2	4	6	8
2	4			
4		16		
6			36	
8		32		48

(c)

-	5		7
	10	6	12
4	9	5	11
	8	4	10

(d)

÷		
9	36	18
6	24	12
2	8	4

Figure 9.1

2. Have child list all of the basic facts that sum to 5, 12, and 18. Note if child needs concrete representation as an aid.

    Example
    $$5 = 5 + 0, 4 + 1, 3 + 2, 0 + 5, 1 + 4, 2 + 3$$

3. List all basic multiplication facts for 36.

    Examples

    $$4 \times 9, 6 \times 6, 9 \times 4$$

4. List all basic division facts for 36.

    Examples

    $$36 \div 4 = 9, 36 - 6 = 6, 36 \div 9 = 4$$

5. Complete relation diagrams such as the following.

(a)   Add 2

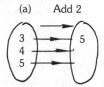

(b)   Multiply by 3

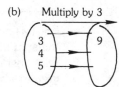

Teaching Activities

The following instructional strategies are based on the assumption that the child has demonstrated a knowledge of the number concepts 0–9 and of the binary operations that relate the numbers.

> ### Instructional Strategy 9.1.1:
> *Emphasize patterns.*

Patterns are particularly appropriate for helping one to see which numbers are components of a larger number. Patterns may be shown at both the concrete and picture levels. Examples of patterns are shown in Figures 9.2 and 9.3. Included are playing cards, dominoes, and grids.

> ### Instructional Strategy 9.1.2:
> *Use rehearsal strategies.*

Rehearsal strategies may be visual or auditory. Visual imagery rehearsal strategies include picturing a pattern, such as those shown in Figures 9.2 and 9.3. Kellas, Ashcroft, and Johnson (1973) had retarded children use cumula-

139

tive rehearsal—"two plus three equals five"; "two plus four equals six." Encourage the child to repeat this to himself.

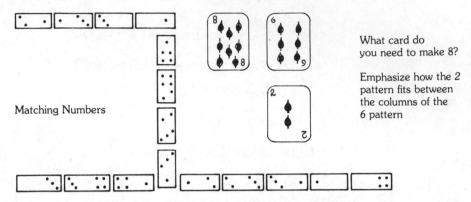

Matching Numbers

What card do you need to make 8?

Emphasize how the 2 pattern fits between the columns of the 6 pattern

Figure 9.2.   *Playing Cards and Dominoes to Show Patterns*

> **Instructional Strategy 9.1.3:**
> *Emphasize differences in distinctive features of stimuli.*

For basic multiplication facts with the number 9 as one factor, emphasize that the sum of the digits of any multiple of 9 is 9. Another distinctive feature of the relationship among basic facts involving 9 is that in the table of nines (for facts greater than $0 \times 9 = 0$), the tens digit *increases* by one as the units digit *decreases* by one (see Figure 9.4).

This is an example of how this IS may be used as a check for correctness of products. Other applications may serve to highlight such distinctive features as form, color, function, and so on when constructing higher-level generalizations. For example, "all round things roll," "red on traffic lights means *stop*," "sharp edges cut."

It is important that the teacher develop patterns such as this with caution. Unless the child is constructing the pattern, he or she may become confused with such patterns or shortcuts. The student should not be left with a jumble of short rules that he or she cannot tie to a foundation. Some use of patterns is helpful, but do not become carried away stressing these to a student.

> **Instructional Strategy 9.1.4:**
> *Restrict size of sum.*

Use sets as a model for investigating basic addition facts. Give the child as many objects as he can comfortably enumerate, up to nine. Then ask him to make as many combinations as he can by placing the elements in two piles (perhaps on two sheets of paper). For example, given six elements (see Figure 9.5), you may have the child record his observations. Encourage him to point up patterns of order and sequence by asking the following questions:

Pattern of Multiples of Five

1	2	3	4	5	6	7	8	9	10	11	12
2	4	6	8	10	12	14	16	18	20	22	24
3	6	9	12	15	18	21	24	27	30	33	36
4	8	12	16	20	24	28	32	36	40	44	48
5	10	15	20	25	30	35	40	45	50	55	60
6	12	18	24	30	36	42	48	54	60	66	72
7	14	21	28	35	42	49	56	63	70	77	84
8	16	24	32	40	48	56	64	72	80	88	96
9	18	27	36	45	54	63	72	81	90	99	108
10	20	30	40	50	60	70	80	90	100	110	120
11	22	33	44	55	66	77	88	99	110	121	132
12	24	36	48	60	72	84	96	108	120	132	144

Pattern of Multiples of Four

1	2	3	4	5	6	7	8	9	10	11	12
2	4	6	8	10	12	14	16	18	20	22	24
3	6	9	12	15	18	21	24	27	30	33	36
4	8	12	16	20	24	28	32	36	40	44	48
5	10	15	20	25	30	35	40	45	50	55	60
6	12	18	24	30	36	42	48	54	60	66	72
7	14	21	28	35	42	49	56	63	70	77	84
8	16	24	32	40	48	56	64	72	80	88	96
9	18	27	36	45	54	63	72	81	90	99	108
10	20	30	40	50	60	70	80	90	100	110	120
11	22	33	44	55	66	77	88	99	110	121	132
12	24	36	48	60	72	84	96	108	120	132	144

Pattern of Adding Two

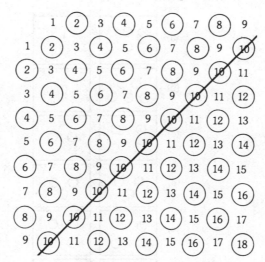

Figure 9.3.   *Grids to Show Patterns*

What do you notice about the two parts?

If you know that 2 + 4 equals 6, what other combination do you also see?

(6 + 0, 5 + 1, 4 + 2, 3 + 3, 2 + 4, 1 + 5, 0 + 6)

If appropriate to the child's level of comprehension, you may wish to point out the commutative property for addition that becomes apparent from the overall pattern.

141

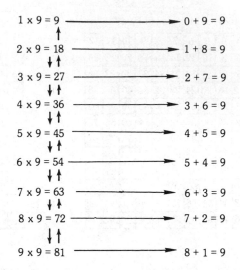

1 x 9 = 9	→ 0 + 9 = 9
2 x 9 = 18	→ 1 + 8 = 9
3 x 9 = 27	→ 2 + 7 = 9
4 x 9 = 36	→ 3 + 6 = 9
5 x 9 = 45	→ 4 + 5 = 9
6 x 9 = 54	→ 5 + 4 = 9
7 x 9 = 63	→ 6 + 3 = 9
8 x 9 = 72	→ 7 + 2 = 9
9 x 9 = 81	→ 8 + 1 = 9

Figure 9.4.   *Basic Multiplication Facts with Nine as a Factor*

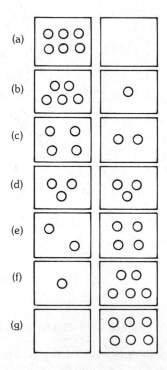

Figure 9.5

Instructional Strategy 9.1.5:
Use R-K counting board.

Use an R-K counting board[1] to investigate basic facts for addition. The basic board has nine spaces with increasing values of one through nine for each space, as shown in Figure 9.6.

9
8
7
6
5
4
3
2
1

Figure 9.6

An empty board has a value of zero. Be sure that the child realizes that a value of one is obtained by "jumping onto the board," at position or space *one*. This is important, because moves that represent "counting on" as a model for addition of basic facts must start from the zero position (off the board). (For example, see 3 + 5 = 8, shown in Figure 9.7.)

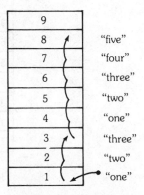

Figure 9.7

Tell the child the value of each space. Practice counting the spaces. Move a token up the board, and identify its value as determined by its position on the board. A digit may be written on each space to cue its value. Use only one token initially with a rule that the board's value is shown by the token's position. Figure 9.8 shows some sample boards.

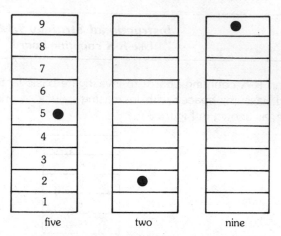

five   two   nine

Figure 9.8

To represent 3 + 2, place the token on space three (or count up to space three from off the board, as shown in Figure 9.7) and then count up two additional spaces with your finger. Move the token to this space, which has a value of five. Work with the child on basic facts with sums to nine until she is adept at using the board. This serves to drill on basic addition facts.

Basic subtraction facts may also be investigated on the R-K counting board. For example, to show 7 – 4 = 3, place the token on the space whose value is seven and count backward four spaces to three, as shown in Figure 9.9.

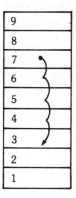

Figure 9.9

Basic multiplication facts to 8 may be shown, as in Figure 9.10. How many 4's make 8? The child counts four spaces, then four spaces again. He is asked how many times he counted four spaces to get to space 8. He answers "twice" or "two times." Record this as

    "2 fours ⟶ 8."

After a while translate this to the standard notation

    $2 \times 4 = 8$.

Again, entering the board by jumping onto space one must be emphasized.

144

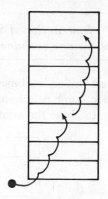

Figure 9.10

Basic division facts to 9 may be shown also. How many 3's can you get out of 9? Start at 9 and count down the board by steps of three (see Figure 9.11).

$$
\begin{array}{r}
9 \\
-3 \\
\hline
6 \\
-3 \\
\hline
3 \\
-3 \\
\hline
0
\end{array}
\qquad
3\ \overline{)\ \begin{array}{r} 9 \\ -3 \\ \hline 6 \\ -3 \\ \hline 3 \\ -3 \\ \hline 0 \end{array}}
\begin{array}{l}
1 \\
\\
1 \\
\\
1 \\
\hline
3
\end{array}
$$

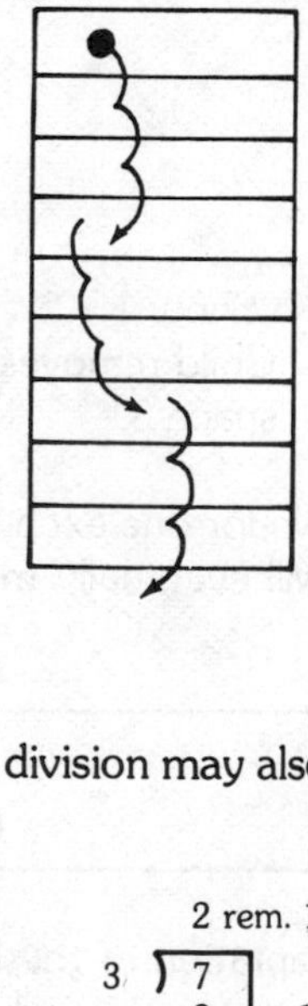

Figure 9.11

For more advanced students, uneven division may also be shown. Take
7 ÷ 3 = 2 remainder 1 (see Figure 9.12).

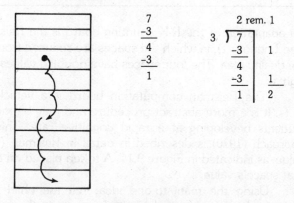

$$
\begin{array}{r}
7 \\
-3 \\
\hline
4 \\
-3 \\
\hline
1
\end{array}
\qquad
3\ \overline{)\ \begin{array}{r} 7 \\ -3 \\ \hline 4 \\ -3 \\ \hline 1 \end{array}}
\begin{array}{l}
\\
1 \\
\\
1 \\
\hline
2
\end{array}
\qquad \text{2 rem. 1}
$$

Figure 9.12

The remainder of 1 is represented by the token remaining on space 1.

> ## Instructional Strategy 9.1.6:
> ### Make two-for-one exchanges on R-K counting board.

After a child has been drilled on basic addition facts with the R-K counting board, show him how to combine tokens representing two addends and make a two-for-one exchange to show the sum. This is demonstrated in Figure 9.13.

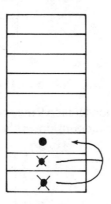

Figure 9.13

"Show 1 + 2 = 3 making a two-for-one exchange."
Child removes the tokens on spaces 1 and 2, and places one token on space 3.

This two-for-one exchange is a forerunner of the ten-for-one exchange that the child will eventually make when generating place values (discussed in section 9.3.0).

> ## Instructional Strategy 9.1.7:
> ### Use Reisman computation board.

An adaptation of the R-K counting board is the *Reisman computation board* (see Endnote 1), in which the spaces are reduced from nine to four by utilizing the duality idea. The four spaces have doubled values (ones place, twos, fours, eights).

The Reisman computation board is a vehicle for investigating basic facts. It is a more abstract procedure and may be more appropriate for those students developing at a rapid cognitive rate. This modification of Papy's approach (1970) is described in detail in Reisman (1977). Each space has a value, as indicated in Figure 9.14. A token placed on a space represents *one* of that space's value.

Using the many-to-one idea, tell the child to make a two-for-one exchange in order to move from one space to the next higher. In making the two-for-one exchange, some children cannot go directly from space one to

Figure 9.14

space two. They need to exchange their two "space one" tokens for a third token and then place that on space two.

Tell the child to identify all the ways he can show a number on a board. You might wish to have several boards available so that all combinations of a number may be shown. Figure 9.15 shows the number 4 in different ways. Note that the basic fact which is not apparent, 3 +1, may be obtained by applying the associative or group property:

$$[(1 + 1) + 1] + 1 \text{ or } (2 + 1) + 1.$$

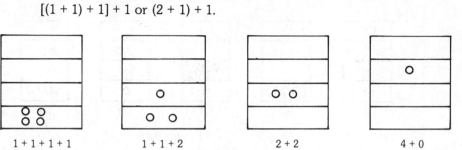

| 1 + 1 + 1 + 1 | 1 + 1 + 2 | 2 + 2 | 4 + 0 |

Figure 9.15

Basic multiplication facts can be shown by using the model of union of disjoint equivalent sets. What basic multiplication facts comprise the number 12? (See Figure 9.16.)

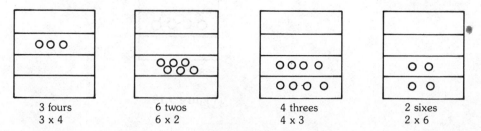

| 3 fours | 6 twos | 4 threes | 2 sixes |
| 3 x 4 | 6 x 2 | 4 x 3 | 2 x 6 |

Figure 9.16

Basic subtraction facts, employing the take-away idea, may be shown as follows: Remove those shown as crossed out (X) from the board. Tokens or checkers may be used. (See Figure 9.17.)

Remember, the child may have to make some exchanges so that the subtraction is readily apparent on the board. For example, in 9 – 7 = 2 the child may start by showing nine on the board as 8 + 1. The number to be subtracted

147

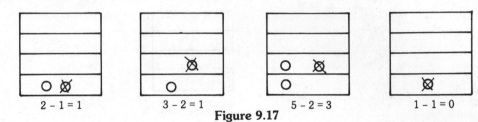

Figure 9.17

is not apparent. Thus, the child may have to be directed to show nine another way. Some children will now use a planned strategy and show nine as

4 + 2 + 1 + 2;

others may take longer, using trial-and-error. Encourage children to show a separate board for each name for nine. Ask, "Which boards show 9 – 7 = 2?" (See Figure 9.18.) 9 – 7 is apparent on boards b, c, e, and f.

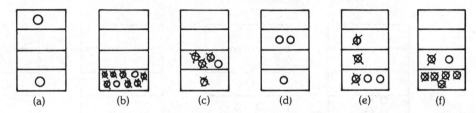

Figure 9.18

For division facts, ask, "How many twos can you get out of 8?" This is based on a repeated subtraction model for division and is shown in Figure 9.19.

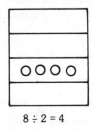

8 ÷ 2 = 4

Figure 9.19

9.2.0 Axioms

There are 11 basic properties of the *real number system*, listed in Appendix A, that underlie the developmental mathematics curriculum upon which this book is structured. The following assessment activities tap a student's knowledge of these axioms.

148

**Assessment
Activities:**

1. Identify the following properties by matching.

____ a. $3 + 5 = 5 + 3$
____ b. $4 + 0 = 4$

____ c. $3 (60 + 5) = 3 \times 65$

____ d. $7 + -7 = 0$
____ e. $7 \times 2 = 2 \times 7$
    f. $4 \times 1/4 = 4/4$

____ g. $7 \times 1 = 7$
____ h. $2 + (3 + 6) = (2 + 3) + 6$

____ i. $(2 \times 7) \times 4 = 2 \times (7 \times 4)$
____ j. the sum of even numbers is an even number
____ k. the product of odd numbers is an odd number

1. multiplicative identity
2. distributive property of multiplication over addition
3. closure for addition of even numbers
4. multiplicative inverse
5. associativity for addition
6. commutativity for multiplication
7. additive identity
8. closure for multiplication of odd numbers
9. additive inverse
10. commutativity for addition

11. associativity for multiplication

2. Complete the following:

a) $\square + \triangle = \triangle +$ _____

b) $\square (\triangle + O) = \square \triangle +$ _____

c) $(\square + \triangle) + 0 = \square + (\triangle +$ _____$)$

d) $\square \times \triangle = \triangle \times$ _____

e) $n +$ _____ $= 0$

f) $n \times 1 =$ _____

g) $(O \times \square) \times \triangle = O \times (\square \times \triangle)$

h) $n \times 1/n =$ _____

i) $n + 0 =$ _____

Teaching Activities

Instructional Strategy 9.2.1:
Emphasize patterns.

Investigation of patterns within a grid illustrates many of the axioms shown previously in the Assessment Activities.

+	0	1	2	3	4
0	0	1	2	3	4
1	1				
2	2				
3	3				
4	4				

additive identity
$(n + 0 = n)$

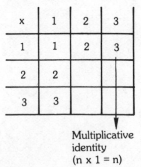

Multiplicative
identity
$(n \times 1 = n)$

149

Instructional Strategy 9.2.2:
Point out relevant relationships.

Investigate relationships within a grid. Cells g and j in Figure 9.20 illustrate the commutative law for addition:

$(2 + 1 = 1 + 2)$

Cells j and 0 illustrate the associative property for addition:

$$(2 + 1) + 2 = 2 + (1 + 2)$$
$$3 + 2 = 2 + 3$$
$$5 = 5$$

+	0	1	2	3
0	a	e	i	m
1	b	f	j3	n
2	c	g3	k	o5
3	d	h	l	p

Figure 9.20

The commutative law for multiplication is shown in Figure 9.21.

$5 \times 3 = 3 \times 5$

Associative law for multiplication:

$$(3 \times 3) \times 5 = 3 \times (3 \times 5)$$
$$9 \times 5 = 3 \times 15$$
$$45 = 45$$

x	3	5	9
3	9	15	
5	15		45
9	27	45	
15	45		

Figure 9.21

(IS)

> **Instructional Strategy 9.2.3:**
> **Use modified conservation task.**

Ask child to show all of the mathematical laws (properties, axioms) he or she can, given sets of five blocks as shown in Figure 9.22, and to identify each.

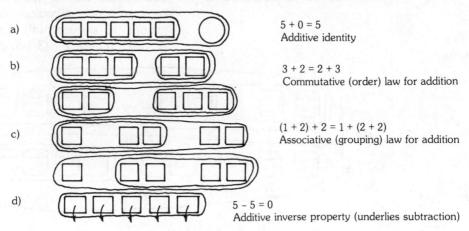

a) $5 + 0 = 5$
Additive identity

b) $3 + 2 = 2 + 3$
Commutative (order) law for addition

c) $(1 + 2) + 2 = 1 + (2 + 2)$
Associative (grouping) law for addition

d) $5 - 5 = 0$
Additive inverse property (underlies subtraction)

Figure 9.22

This approach is particularly appropriate for the gifted child in the primary grades. Also, for the creative child developing at a slower cognitive rate, this strategy allows for divergent thinking, which is a strength of a creative problem solver. Such a child can be a detective and "uncover" hidden laws by investigating the situation.

The laws that apply to multiplication may also be investigated in this manner, as follows:

Multiplicative identity: Arrays in the form shown in Figure 9.23 illustrate the idea that *any* number times one is that number, or that one of any number is that number regardless of materials used or linear spatial arrangement.

1 x 5

5 x 1

Figure 9.23

Commutative law for multiplication: Regardless of materials used, except liquid, child states that "I have 15 pieces of candy if there are three bags with five candies, or five bags with three candies."

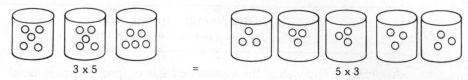

3 x 5   =   5 x 3

Associative law for multiplication: Child states that he has a liter of milk or any liquid regardless of how he groups 10 deciliter (dl) containers— (2 + 3) + 5 dl or 2 + (3 + 5) dl.

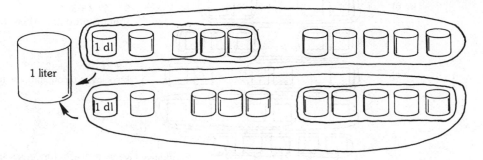

9.3.0 Place Value

Place value allows a numeration system with a finite number of digits to express numbers greater than the face values of the system's digits. For example, we use a finite number of 10 digits, 0–9, but we can represent numbers of greater value by reusing these same digits and employing place value.

When a child is learning how to write numerals in a counting sequence—such as 0, 1, 2, 3, or 9, 10, 11—what appears simple is really complex. Writing numerals (digits are one-place numerals) in a "plus one" sequence involves an *add-one relation*. This is a model for addition, and the result is a *sum*; for example,

$$0 + 1 = 1; 1 + 1 = 2;\quad 2 + 1 = 3; \ldots 8 + 1 = 9.$$

The sequence of written numerals from 9 to 10 is a key move. Children initially consider the names "ten," "eleven," and "twelve" in the same class of labels as "one," "two," and so on up to "nine." However, because the move from 9 to 10 involves attention to place value, care must be taken to distinguish between instruction that focuses on sequencing and procedures that involve making a many-for-one exchange. Many children can accommodate these two components of place value numeration through incidental learning. Children with special learning needs who do not notice salient aspects of situations *must be helped* to bridge from the add-one sequence of writing numerals in order to the many-to-one generalization that creates place values.

Many-to-one is a lower-level generalization that is a foundation of the multiplication operation. Thus, when a child writes

9, 10, 11, 12,

he may not be thinking "nine, next I have ten units that I will exchange for one ten, so I have one ten and no units; then I have one ten and one unit; then one ten and two units." He is more likely thinking, "nine and one more is ten," and writes *10* because he has learned by rote to associate the numeral *10* with the word *ten. Sequencing by adding one to nine to obtain ten is not the same as making a ten-for-one exchange to get one ten.*

Place value is the *product* obtained by multiplying a numeration system's base—the product of powers of the base. Considering the units or ones place as the center of the system in base ten, we have:

$$10 \times [10 \times (10 \times 1)], 10 \times (10 \times 1), 10 \times 1, 1/10 \times 1, 1/10 \times (1/10 \times 1),$$

and so on. Thus, the thousands, hundreds, tens, tenths, hundredths place values are products.

Multiplication is also involved when we determine the value of a numeral in another way: face value times place value. Thus, the value of each digit in a numeral is a product; for example,

$$2 \text{ tens} = [2 \times (10 \times 1)].$$

For some reason, the multiplication ideas that a child encounters in his or her experience are not pointed out until second or third grade, but both *generating place values* and *face value times place value*—which are products—are taught in grade one.

Now a word about bundling sticks to teach place value. Place value is a notational convention, while bundling objects in groups of 10 is a number idea. Bundling sticks of 10 is not a physical embodiment of the add-one relation and should not be taught as though it were. The bundling procedure provides the multiplication idea underlying powers of 10, and therefore it is a prerequisite for expressing numbers in base 10.

There are, however, three important discrepancies between our notational system and grouping objects to represent base 10 numbers. First, in order to represent the null or empty set in our notational system we use *something*—the digit 0. But in order to physically represent the empty set, we use nothing. This points up the difference between the number 0 and the symbol for zero.[3] Second, as stated previously, generating base 10 place values using powers of 10 is not the same as counting in sequence, which is an add-one relation. For this reason, bundling sticks of 10 to form one 10 may be confusing. Third, we use *two* digits to represent one 10, yet *one* bundle of sticks is supposed to represent a two-digit numeral. Also confusing is the widely used "place value chart," in which 10 sticks are inserted into the units partition, then bundled with a rubber band and moved to the tens compartment. We may not write more than a single digit in a particular position. Inserting several sticks into the units compartment of the place value chart does not match the rule of notation that allows only *one digit* per place. Therefore, when using the bundling method, be sure to exchange the 10 unit sticks for a single stick which represents one 10.

In the early stages of developing our notational system, a "counting-on" model seems to resemble more closely how numerals are written in a counting sequence rather than a "grouping by ten" model. This distinction in procedure is shown in Instructional Strategy 9.3.1 using the R-K counting boards.

Assessment
Activities:

1. Write the following in expanded notation.
   A. $32 = (3 \times \Box) + (2 \times 1)$
   B. $46 = (\Box \times \Box) + (\Box \times \Box)$
   C. $723 =$
   D. $309 =$

2. Write the following in standard form—for example, $(7 \times 10) + (6 \times 1) = 76$.

   A. $(3 \times 10) + (2 \times 1) = 3 \,\_\_$
   B. $(7 \times 100) + (4 \times 10) + (2 \times 1)=$
   C. $(4 \times 10) + (0 \times 1)$

3. Write the following face values and place values.
   A. $376 = (\Box \times 10) + (\Box \times 1) + (\Box \times 100)$
   B. $72 = (2 \times \Box) + (\Box \times 10)$
   C. $987 = 7$ units, 9 _____, 8 tens

Teaching Activities

Review Trouble Spots II, III, and IV, page 255 in Appendix E, before proceeding.

> **Instructional Strategy 9.3.1:**
> *Use R-K counting boards: "sequence" model for numerals up to 19.*

Engage the child in activities to familiarize him with the R-K counting boards, described in IS 9.1.5. Point out the similarity between the adding-one relation and bridging in sequence from 9 to 10. Increasing values within the units board are represented by the vertical arrangement of spaces, designated by the sequential values 1–9. When the child moves his token up through the board as he counts to 9, and then has no space for 10, point out that there are no more *digits* beyond 9. This presents a problem-solving situation. Questioning may elicit the fact that *two* of the original digits are reused to write the numeral for ten; 10 is a two-place numeral.

Have available a second board with no digits written on it. At this point you may simply tell the child that a second board is needed. Arrange the additional board as shown in Figure 9.24.

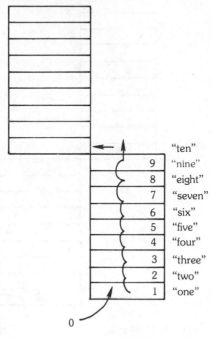

Figure 9.24

The positioning of the "tens board" to the upper left of the units board reinforces the idea of increase in number. Eventually the tens board will be lowered to a position adjacent with the units board to reflect the written form of a multi-digit numeral.

The need for the new board should be shown to correspond to the need for a new place value position in which to write a second digit representing each count to 10. When the token jumps off the top of the units board, a decision is made to move it onto space *one* of the newly added board. Since the count at this point is *ten*, and the movement to the left is lateral, thus indicating no change in value, the value of the bottom space on the new board is seen to be *ten*.

Using one token, guide the child to count from the ones space on the units board up to the ones space on the tens board. The important bridge is that the child sees the relationship of the picture to the numeral 10. The token represents one 10, and the empty units board shows a value of zero. You may also wish to record the digit below each board to represent number value.

From *ten*, introduce a second token for the units board to show *eleven*. A common mistake occurs at this point. Many children, when asked to show *eleven*, move the token from the tens place to the twenties space. This error may be avoided by (a) placing a second token in the ones space on the units

155

board and (b) counting "eleven" as the token "jumps onto" the ones space (see Figure 9.25). Have the child practice this procedure until he can use the boards and record a number to 19 with 100% accuracy.

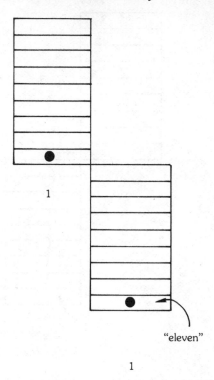

1

"eleven"

1

Figure 9.25

Instructional Strategy 9.3.2:
Use R-K counting boards for writing numerals greater than 19.

This IS bridges the sequencing procedure to the exchange model. Move the units board token upward as the count progresses from 19. At 20, the units board token moves over to the bottom space on the tens board. Establish a ground rule that calls for the fewest possible tokens to show a number; then make a two-for-one move upward to the twenties space. (See IS 9.1.6 to review the two-for-one exchange.)

The two-for-one move on the tens board, which involves exchanging two 10s for one 20, involves the same reasoning as exchanging 10 units for one 10. The child must understand that the two-for-one transformation does not affect the equivalence of the two conditions; the identity of the number, 10, is preserved across the transformation.

> ## Instructional Strategy 9.3.3:
> *Use R-K counting boards: Exchange model for writing numerals greater than 9.*

Arrange two R-K boards as in Figure 9.26, and review how the second board provides for showing numbers greater than 9. Remind the children about the two-for-one moves. Set two new rules, stating that this time (a) two-for-one moves must equal the number 10, and (b) they can temporarily overload the units board. To overload the units board means that the children can place two or more tokens on it at a time to show the number 10. They should show 10 as follows on the units board by placing tokens on: 1 and 9, 2 and 8, 3 and 7, 4 and 6, or two tokens on 5. They then exchange an overloaded units board for one token placed on the tens board. Figure 9.26 shows this forward move for the exchange 4 and 6 for one 10.

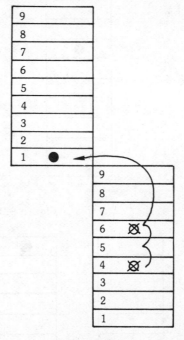

Figure 9.26

Some children may find it helpful to overload the units board by placing 10 tokens on the ones space. They then make many small two-for-one moves upward, until the final 10-units-for-one-10 exchange is represented by one token jumping up off the top of the board and moving laterally to the left, to be placed on the tens space of the tens board.

The relationship between transformation and invariant must be understood if "carrying" and "borrowing" presented below is to have meaning. We shall see how a forward move from the units board to the tens space is the same as renaming in addition. Also, a backward move—exchanging a token in the tens space on the tens board to two or more tokens on the unit boards—is the physical model for renaming in subtraction. The single token on the twenty space, which stands for two tens, is analogous to the face value function of a digit. The digit 2 stands for *two* of something regardless of its position in a digit. Likewise, a token placed in the second space up stands for *two*, regardless of the overall value of each space within a board. Therefore, space value within a board is analogous to face value, and the overall value of a board—depending on its position in relationship to the units board—determines the place value of a digit in a numeral.

At some point, the tens board may be lowered adjacent to the units board. This completes the physical representation of a two-digit or three-digit numeral, as shown in Figure 9.27 for the number 74 and in Figure 9.28 for the number 391.

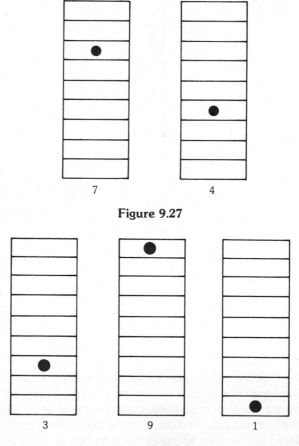

Figure 9.27

Figure 9.28

Notice that each time a movement from the units board to the tens board occurred, there was an equivalent exchange in value of 10 for 10, and the units board was left empty. This happened regardless of whether it resulted from a sequence procedure or from an exchange.

The spaces on the tens board may be thought of as keeping a tally count either of (a) the number of times the count of 10 was made or (b) the number of many-to-one exchanges where the value *ten* was the invariant.

> ### Instructional Strategy 9.3.4:
> ### Use Reisman computation boards.

For children who enjoy a more complex model, use again the Reisman computation boards, described in IS 9.1.7. Make a two-for-one forward move to the tens place to show 2 + 8 as one 10. Eventually, place the units and tens boards next to each other, as shown in Figure 9.29.

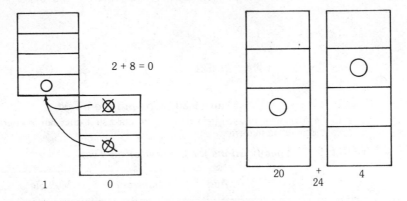

Figure 9.29

9.4.0 Computation

Computation involves algorithms, which are procedures for operating on numbers, and makes use of one's ability with basic facts, relevant axioms, and place value.

Assessment Activities:

The computations shown in Figure 9.30 served as the test in a physiological measures study (Reisman, Braggio, Braggio, Farr, & Simpson, 1977). Table 9.1 shows the objectives that served as a structure for the computations. The test items in Figure 9.30 are arranged in Table 9.1 by operation, from simple to complex. Basic facts are imbedded in more complicated algorithms so that the teacher may view the effect of increased task complexity on performance.

(1)   7
      +2

(2)   17
      +42

(3)   6
      +7

(4)   36
      +57

(5)   306
      +57

(6)   47
      −12

(7)   56
      −37

(8)   506
      −127

(9)   23
      x3

(10)   23
      x4

(11)   23
      x34

(12)   76
      x38

(13)   2)84

(14)   3)84

(15)   5)84

(16)   3)603

(17)   3)804

(18)   23)4623

(19)   17)1718

Figure 9.30. *Computation Test*

Omit number of each item on copy for child so it does not interfere with computation numerals.

TABLE 9.1. *Specifications for Computation Test*

	Add	Subtract	Multiply	Divide
Basic fact < 10	1			
Basic fact > 10	3			
2 place (no renaming)	2	6	9	14
2 place (renaming)	4	7	10, 11, 12	15, 16
> 2 place (renaming)	5	8	13	18, 19
Zero in numeral	5	8	13	17, 18
No remainder				14, 15 17, 18, 19
With remainder				16, 20
> 2 place (no renaming)				17, 20

160

Teaching Activities

Review Trouble Spots VII and VIII, pages 256–257 in Appendix E, before proceeding. Also, review section 9.1.0 of this chapter on basic facts.

```
Instructional Strategy 9.4.1:
Use R-K counting boards.
```

Extend the rules to allow overloading of a board and use of more than one token. For example, in adding basic facts with sums greater than 9, show the addends 3 and 9 each with a token, as shown in Figure 9.31.

9	●
8	
7	
6	
5	
4	
3	●
2	
1	

Figure 9.31

There are two computation procedures that are helpful for those who have not learned to compute from conventional methods.

Procedure 1

For the example 9 + 3, the token representing the smaller addend (3) is replaced by three counters, as shown in Figure 9.32.

The counters are positioned as follows: one space atop the units board, next to the ones space, and one next to the twos space. These counters then serve as cues for obtaining the sum (12). The child removes the nines-space token and counts the counters in sequence: "ten," "eleven," "twelve." The child next shows 12 on the boards by moving (a) the counter positioned next to the top of the board to the left onto the tens space within the tens board and (b) the counter he called "twelve" onto the twos space of the units board. The child may then place cut-out numerals below each board, or he may write them on a piece of paper placed below the board.

Now let us use the same procedure to compute 8 + 5. Replace the smaller addend (5) by five counters, as shown in Figure 9.33.

The child removes on the eights space and counts the counters in sequence up off the top of the units board and then beginning next to the ones space, up through the units board again, as shown: "nine," "ten," "eleven," "twelve," "thirteen." Show 13 on the boards, as indicated by the solid tokens.

161

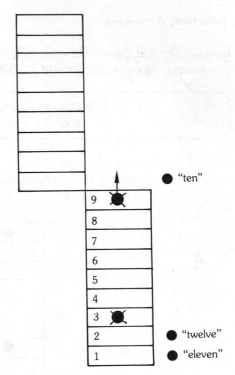

Figure 9.32

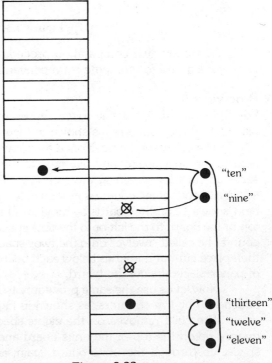

Figure 9.33

Procedure 2

For the problem

$$3 + 9 = 12,$$

proceed as shown in Figure 9.34. Employing a two-for-one forward move, rename the smaller addend so that 10 is apparent on the overloaded units board. Then make a two-for-one exchange onto the tens board:

1. Exchange the three-space token for a two-space and a one-space token.
2. Make a two-for-one exchange (9 + 1) making 10, and show this with a token on the tens space.
3. The boards show one 10 and two units, or *12*.

For *5 + 8 = 13*, the procedure is shown in Figure 9.35.

1. Show 5 as 3 + 2.
2. Combine 8 + 2 to make 10. This constitutes a two-for-one exchange onto the tens space on the tens board.
3. The board shows one 10 and three units.

Procedure 2 is a model for "carrying," or renaming, in addition. Point out that there is no position on the units board to show 13 with a single token.

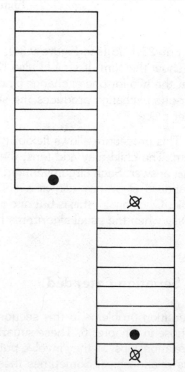

Figure 9.34

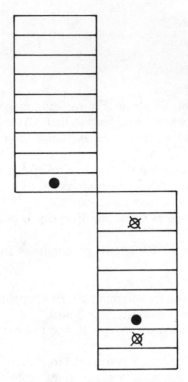

Figure 9.35

For *23 + 35* (see Figure 9.36), the two-for-one exchanges within each board show the sum. It should also be noted that the sum may be obtained without the two-for-one exchange by counting as follows: "30, 50, 55, 58." The two-for-one exchange produces the sum in standard form, that is, using one digit per place.

This procedure allows flexibility in the sequence of computing within a problem. The child may add tens, then units, then "carry," then add tens for the final answer. Such circumvention, although cumbersome in relationship to the standard algorithm, allows a child to use his own sequence within the example. Once again, this is but one procedure of many. This should be tried, especially when the usual algorithms have met with failure.

9.5.0 Seriation Extended

The seriation problems in this section involve patterns of greater complexity than those in Chapter 6. These situations go beyond simple generalizations of basic relationships, as they involve putting two or more concepts together into various relationships. Sometimes these relationships are quite subtle and the

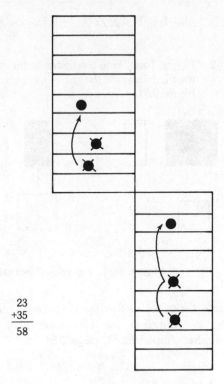

$$\begin{array}{r} 23 \\ +35 \\ \hline 58 \end{array}$$

Figure 9.36

underlying rule is not readily apparent. Series problems were described as follows:

> The principle of a series can be grasped only if one perceives the collection of numbers as a whole, without forgetting the individual numbers in a series. From a strictly mathematical standpoint the problems in the tests below can be solved in many ways. (Krutetskii, 1976, pp. 151–152)

Assessment Activities:

1. Numerical Test (In parentheses are the expected numbers to continue series.)
    - A. 2; 4; 6; 8; 10   (12)
    - B. 19; 16; 13; 10; 7   (4)
    - C. 3; 9; 27; 81; 243   (729)
    - D. 1; 4; 9; 16; 25   (36)
    - E. 1; 2; 4; 7; 11   (16)
    - F. 10; 8; 11; 9; 12   (10)
    - G. 2; 3; 4; 9; 16   (81)
    - H. 1; 2; 6; 24; 120   (20) multiply preceding number by 2; 3; 4; etc.
    - I. 2; 4; 10; 28; 82   (244) multiply preceding number by 3 and subtract 2.

J. 1; 5; 14; 33; 72   (151) double preceding number and add 3; 4; 5; etc.

2. Figural Test: "Find the figure on the right which would be related to the third figure on the left as the second is related to the first." (See Figure 9.37.)

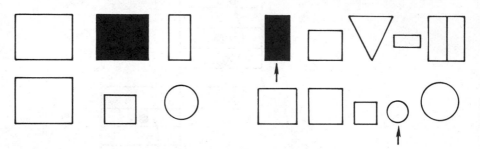

Figure 9.37. *Extended Seriation*

Additional tasks for assessing a student's ability to perform complex seriations may be found in the Developmental Mathematics Curriculum Screening Checklist, Appendix D, page 251.

Teaching Activities

The number tasks demand that the student perceive the entire collection while simultaneously considering relationships between individual numbers. The tasks in this section are appropriate for students developing at a rapid cognitive rate. However, Krutetskii (1976, p. 151) pointed out that "the level of mathematics development of even the ablest pupils in grades five through seven does not prepare them to perceive the different principles that might underlie a given series."

> *Instructional Strategy 9.5.1:*
> *Provide necessary structure.*

Do not assume that just because a child is gifted she intuitively "knows" all mathematics. Some children need to be *led* into a task to get them started. Structure may take several forms. The teacher may think aloud while solving some extended series tasks, thus serving as a model; brainstorming sometimes helps; allowing children to team-solve series is helpful.

> *Instructional Strategy 9.5.2:*
> *Use student's questioning skills.*

Encourage the student to question different relationships across the series. This helps the student to develop an independent problem-solving strategy.

> ### Instructional Strategy 9.5.3:
> *Provide both verbal-logical and visual-pictorial tasks.*

Some students may be able to find the generalization or principle that underlies a numerical series but have difficulty with spatial arrangement of elements. Therefore, provide both types of tasks. This information may be used for identifying the student's strengths and weaknesses. The student who is less able to do numerical sequences may be displaying syntactic disability and may have difficulty in algebra. Those who cannot perform the figural tasks may have difficulty with geometry, which is spatial. Teaching to a student's strengths is important in curriculum selection. If your purpose is for the student to generate series, then be sure the tasks involve examples that relate to his strength.

9.6.0 Solving Word Problems

In section 8.4.0 of Chapter 8, the emphasis was on temporal relationships and their effect upon word problems. Other relationships involved were plurals and singulars, and spatial relationships, especially those expressed in prepositions. In this chapter, the focus is on word problems that express higher-level generalizations due to their complex nature. Each of the following examples expresses a relationship between two or more concepts.

The steps in solving mathematical problems, described by Krutetskii (1976), may be applied to solving word problems. Three basic steps include the following (Krutetskii, 1976, p. 98):

1. gathering the information needed to solve the problem
2. processing this information while solving the problem
3. retaining in one's memory the results and consequences of the solution

Assessment Activities:

1. A boy has to build a dog house for his two dogs. He needs an inside measure of 2 cubic meters. If the house is to be 1 meter wide and 1 meter deep, how high must he build it?

2. A man has 300 apple trees. If each tree yields 35 apples a season, how many apples does the man produce each season?

3. Fourteen children want to divide a pound of gum drops. If there are 42 gum drops, how many will each child get?

4. A woman has 4 liters of milk. She needs 6½ liters. How much more milk does she need to get?

5. Three boys and two girls have three candy bars. If the boys divide them equally among themselves, how many candy bars will each girl receive?

6. A girl buys 28 meters of cloth to make curtains. If each curtain takes 5⅜ meters, how much more material does she need to make three pairs of curtains?

Teaching Activities

The following activities are suggested for rapid mathematics learners.

> ### Instructional Strategy 9.6.1:
> ### *Present problems with an unstated problem.*

Present "problems where the question is not stated, but proceeds logically from the mathematical relations given in the problem" (Krutetskii, 1976, pp. 105–107). Examples follow of such arithmetic, algebra, and geometry tasks (taken from Krutetskii, 1976, pp. 105–107); the unstated question or task is in parentheses.

Arithmetic Tasks

Twenty-five pipes of length 5m and 8m were laid over a distance of 155m. (How many pipes of each kind were laid?)
A boy has as many sisters as brothers, and his sister has half as many sisters as brothers. (How many brothers and how many sisters are in the family?)

Algebra Tasks

A pupil bought $2b$ notebooks at one store and 3.5 times as many at another. (How many notebooks did he buy altogether?)
A man has lived a months. (How many years old is he?)

Geometry Tasks

One angle of a triangle is 30° larger than a second, and the third angle is 20° smaller than the first. (Find the size of the first angle.)
In a rectangle the diagonals intersect at a point that is 6 cm farther from the smaller side than from the larger. The perimeter of the rectangle is equal to 44 cm. (Find the length of the sides.)

> **Instructional Strategy 9.6.2:**
> *Present problems with incomplete information.*

"These problems indicate whether and how pupils *catch on* to the formal structure of a problem as they perceive the problem's terms and whether they are able to detect incomplete information" (Krutetskii, 1976, p. 107). These tasks include arithmetic and geometry items. The student is asked the following questions:

> Why can't you give a precise answer?
> What do you lack?
> What must be added?
> Prove that now the problem can be precisely solved.
> Can you extract something even from this information? (For the more capable.)
> What conclusion can be drawn from analyzing what is given? Let the answer be imprecise and indefinite. (For more capable.)

Arithmetic Tasks

A train consists of tank cars, freight cars, and flatcars. There are 4 fewer tank cars than flatcars, and 8 fewer tank cars than freight cars. How many tank cars, freight cars, and flat cars does the train have? (Their total number is unknown.)

A jar of honey weighs 500 g. The same jar weighs 350 g when filled with kerosene. How much does the empty jar weigh? (Need a relationship between the weight of the honey and the weight of the kerosene; for example, kerosene is twice as light as honey.)

Geometry Tasks

Calculate the side of a square of area 64 cm². Calculate the side of a rectangle with area 36 cm². (In the second calculation, either the length of one of the sides or a relationship between the lengths of the sides is needed.)

The sides of a triangle are in the ratio 5:4:3. Find the length of its sides. (Need the length of the perimeter or at least of one of the sides.)

> **Instructional Strategy 9.6.3:**
> *Use problems with surplus information.*

Supplementary or unnecessary cues (shown in italics) are introduced that mask the facts needed for solutions.

In a store, 24 sacks of potatoes weigh 3 kg and 5 kg each, *with more* in the former than in the latter. The weight of all the 5-kg sacks was equal to the weight of all the 3-kg sacks. How much did each weigh?

Given an isosceles triangle with one side 2 cm, another 10 cm, *and the third equal to one of the other two given ones*, find the third side.

> ### Instructional Strategy 9.6.4:
> **Present problems with several solutions.**

Find the sum of all the integers from 1 to 50.

In how many ways can $27 be paid if the money is in $2 and $5 notes? List the various combinations.

> ### Instructional Strategy 9.6.5:
> **Present problems in a series involving several operations.**

This investigates the student's facility in switching from one method of operation to another. Word problems in many basal arithmetic series used to have the same operation for all problems on a page. The students were often unable to solve new word problems in which they had to decide upon the operation because they lacked experience in having to judge the cause–effect relationship between the operation and their answer.

Ask the following questions concerning a series of word problems:

What makes this problem different?

How are these two the same?

What type of answer will you get if you add? Subtract? Will it be too big? Too small? Correct?

For the remaining topics in this chapter, assessment items may be developed from the Developmental Mathematics Curriculum Screening Checklist, Appendix D, as follows:

Checklist		Topic
V-G	9.7.0	*Cause-Effect Extended*
		The ability to generalize cause-effect relationships involves flexibility of thought, awareness of several alternatives to a problem situation, and awareness of consequences of an act. Thinking ahead to the future is also involved.
V-H	9.8.0	*Topological Equivalence*
		Objects that can take on each other's form are equivalent topologically. For example, a square shape can be stretched or twisted to a triangular or circular shape.

Checklist		Topic
V-I	9.9.0	*Projective Geometry*
		Shadows of objects that differ in shape or size of the object in ratio comparisons are projective transformations. For example, one's shadow may be proportionately longer or shorter than the person.
V-J	9.10.0	*Nonmetric Geometry*
		Points, lines, planes, rays, line segments, angles and polygons and their constructions are included under nonmetric geometry.
V-K	9.11.0	*Metric Geometry*
		This topic includes finding the measure of components of topic 9.10.0—for example, length, width, height—including use of metric measure.
V-L	9.12.0	*Network Theory*
		Networks focus on arcs and vertices. A Swiss mathematician, Leonard Euler, found the relationship $F + V = E + 2$ where F represents faces or spaces of a polyhedra, V stands for vertices, E for edges.
V-M	9.13.0	*Probability, Statistics, Graphs*
		Some events are more likely to occur than others. Probability involves the following questions: Are all outcomes equally certain? Are two or more events mutually exclusive? Are two or more events independent?
		Descriptive statistics summarize numerical data; inferential statistics estimate generalizations based on a sampling of a population. Graphs are used to record data; they may show the mean, mode, or median of distributions.
V-N(1–1)	9.14.0	*Applied Mathematics—Measurement*
		This topic includes time, calendar, reading plane and train schedules, clothing size, metric measure.
V-N(5)	9.15.0	*Applied Mathematics—Economics*
		Included here are money values, comparison of prices, budgeting, sales tax, hourly wage, rent, banking, installment buying.

General suggestions for instructional strategies are presented for these final topics. The topics themselves are an instructional strategy, namely, selection of curriculum that is learnable, motivating, and relevant to the student's basic life needs. The following general statements may be useful in selecting ISs already presented for other topics.

Provide activities that give concrete experiences: paper folding (9.8.0, 9.10.0), projecting shadows with a flashlight (9.9.0), work experience (9.13.0, 9.14.0, 9.15.0), making own measuring instruments (9.14.0), making models (9.8.0), scale modeling (9.9.0, 9.14.0).

Provide cues that enhance the underlying relationships and structures: use of color (9.11.0), tracing paths (9.12.0), developing diagrams and graphs (9.13.0).

Use techniques to direct student's attention to salient aspect of situation. For example, Turnbull and Schultz (1979) suggest:

1. Change seating arrangement.
2. Use cues such as touching the student, signaling with the hand with a predetermined and established gesture, or giving verbal prompts, such as "think or "listen."
3. Seat inattentive student near attentive peers who serve as models.
4. Use diagnostic teaching to insure that tasks are commensurate with child's level of understanding and/or performance.

Analyze task to insure that prerequisites have been learned. Make sure directions are understood by using (a) consistent language easily understood by the student, (b) routine formats of seatwork assignments, (c) peer tutor to explain directions to a handicapped student, and (d) the tape recorder to provide a more detailed explanation to the handicapped student.

Use printed materials that have a minimum of distractors for those students who cannot deal with pages with two or three different types of print on one page.

For children with shorter attention spans, provide three 10-minute lessons rather than one 30-minute activity, with intermittent time for the student to regain his energies.

Incorporate various response modes, such as pointing tasks as well as reproduction tasks; speaking as well as writing activities.

Encourage students with a memory deficiency to *use a hand calculator* to circumvent this deficiency.

Allow team taking of tests to reduce levels of anxiety that may be so severe that child's performance is affected.

Avoid teaching splinter skills by structuring mathematics generalizations to everyday living needs; for example, teaching money uses in a classroom store (Shultz, 1973).

Summary Thought Questions and Activities

I. Show counting on the R-K counting boards.
   A. Make a set of boards and with a checker, penny, or token count up to 99.
   B. Show the following numbers on the R-K boards: 73, 47, 103.
II. Compute basic addition facts with sums to 9 on the R-K Counting Board as follows: *1 + 3, 7 + 2, 5 + 4, 4 + 5, 7 + 0.*
III. Compute the following on the R-K boards:
   A.   23
        +46
        ____

   B.   79
        +10
        ____

C. 53
   +92

D. 78
   −13

E. 51
   −31

F. 4
   ×3

IV. List reasons why this procedure (R-K counting boards) is a physical representation of the computations you have just performed. Discuss why this instructional strategy would be appropriate for a middle school or high school student who did not understand place value and its role in these computations.

Suggested Readings

Bartel, N. R. Problems in mathematics achievement. In D. D. Hammill & N. R. Bartel (Eds.), *Teaching children with learning and behavior problems* (2nd ed.). Boston: Allyn & Bacon, 1978.

Cawley, J. F., & Vitello, S. J. Mode for arithmetic programming for handicapped children. *Exceptional Children*, 1972, 39, 101–110.

Chandler, A. M., Immerzeel, G., & Trimble, H. C. (Eds.) *Experiences in mathematical ideas* (Vol. I). Washington, D.C.: National Council of Teachers of Mathematics, 1970.

Goodstein, H. A. The performance of educable mentally retarded children on subtraction word problems. *Education and Training of the Mentally Retarded*, 1973, 8, 197-202.

Peterson, D. *Functional mathematics for the mentally retarded*. Columbus, Ohio: Charles E. Merrill, 1973.

Smith, R. M. Clinical teaching: *Methods of instruction for the retarded*. New York: McGraw-Hill, 1974.

References

Kellas, G., Ashcraft, M. H., & Johnson, N. S. Rehearsal processes in the short-term memory performance of mildly retarded adolescents. *American Journal of Mental Deficiency*, 1973, 77, 670–679.

Krutetskii, V. A. [*The psychology of mathematical abilities in schoolchildren*] (J. Kilpatrick & I. Wirszup, eds., and J. Teller, trans.). Chicago: University of Chicago Press, 1976.

Papy, F. *Les enfants et la mathematique 1*. Bruxelles-Montreal-Paris: Didier, 1970.

Reisman, F. K. *Diagnostic teaching of elementary school mathematics: Method and content*. Skokie, Ill.: Rand McNally, 1977.

Reisman, F. K., Braggio, J. T., Braggio, S. M., Farr, J., & Simpson, L. *Physiological measures of arithmetic performance in children*. Mathematics Information Report, The ERIC Science, Mathematics and Environmental Education Clearinghouse Center for Science and Mathematics, Ohio State University, April, 1977.

Schultz, J. B. Simulation for special education. *Education and Training of the Mentally Retarded*, 1973, 8, 137–140.

Turnbull, A. P. & Schultz, J. B. *Mainstreaming handicapped students: A guide for the classroom teacher*. Boston: Allyn & Bacon, Inc., 1979.

Endnotes

1. For more information concerning the R-K counting boards and Reisman computation boards, contact the authors in care of the publisher.

2. The R-K counting board and the Reisman computation board are only two of the basic devices for building place value learning. There are many more, and no one procedure will work for all students. The R-K algorithms are one alternative.

3. This also shows the difference between set notation for the null or empty set $\{\ \}$ and the set whose element is 0, as shown in $\{0\}$. In the first case the cardinal number property is zero, while in the second situation the cardinality of the set is one.

Instructional Strategies Based on Psychomotor Needs

10

Chapters 5 through 9 presented instructional strategies for mathematics topics within each component of the developmental mathematics curriculum (see page 11). These strategies are appropriate for children developing at cognitive rates either slower or faster than the majority of their age-mates. The assumption was that the sensory modalities were intact, and that no health problem or physical impairment was an influence. This approach allowed us to focus on *cognitive* generic factors that play a part in learning mathematics.

The present chapter focuses on procedures that accommodate instruction for psychomotor disabilities.

10.1.0 Psychomotor Disabilities

Specific learning disabilities are often synonymous with psychomotor disabilities. Terms discussed in Chapter 2 that should be reviewed include the following:

- Visual-spatial organization
- Perceptual integration
- Perception
- Visuomotor disorder
- Psychomotor learning

In keeping with the function approach that is the structure for this book, we define specific learning disabilities in mathematics as having one or more of the following characteristics, to the extent that it creates a problem for the student's performance in mathematics:

Reasoning
- Does not appropriately sequence occurrences or objects
- Displays *logical errors* in computing algorithms; for example, reversing the usual direction while processing an algorithm, or using a form that resembles an historical procedure but that is inconsistent and thus results in a wrong answer
- Inability to make choices and decisions

175

- Does not connect cause–effect relationships
- Does not make appropriate inferences from data, and draws inappropriate conclusions
- Reverses word meanings; for example, big-little, right-left, in-out
- Uses cues incorrectly; for example, in dividing fractions, inverts the dividend and multiplies because focus is on the fact that "one of them gets flipped and then multiply"
- Incorrect use of syntax in language; for example, gets word order confused (sees no difference between "John is lighter than Mary" and "Mary is lighter than John," or states "The boy to the store went")
- Displays receptive language disabilities

Problem Solving
- Does not organize a task to facilitate its completion
- Does not employ efficient problem-solving strategies; often uses random, trial-and-error approach to tasks
- Does not complete work on time; seems to always be in a duel with time

Orientation
- Does not indicate knowledge of spatial relationships; disoriented spatially
- Does not show awareness of temporal relationships; for example, always late for the bus, school, or appointments in general
- Displays reversals and mirror images in writing digits, place value notation, alphabet, words (such as *was-saw*); this occurs beyond a reasonable age— usually about eight. Computes +, ×, – algorithms in left–right direction

Motor Performance
- Clumsy for age
- Excess tongue movement and/or salivation while writing mathematics computations
- Excess body movement, especially if sitting on a movable chair and when mathematics task becomes more difficult for child
- Writing is so messy that mistakes in computation result from incorrect placement or mistaken identity of digits
- Immature eye–hand coordination
- Good gross motor but poor fine motor coordination
  Displays expressive language disabilities

Attention
- Distractible
- Attention flips in and out

Perception
- Displays visual perception problems
- Displays auditory perception problems

176

Affect

- Is often ambivalent regarding self-concept; positive toward self in areas of competence—drawing, painting, sculpting, music, cooking, creative writing, story-telling—but acts defeated in mathematics situations
- Displays inappropriate level of anxiety in relationship to task difficulty[1]

Assessment Activities:

These activities have been developed from behaviors described in Chapter 2.

1. Present two similar visual stimuli and ask the student to point to or circle the one indicated. For example, present the digit 3 and the capital letter E and say, "Circle the *three*."

2. Present a sheet with several digits and ask the student to trace the one indicated. For example, ask the child to trace all of the 2's on the following sheet:

6	2	0
4	3	7
2	8	9
4	3	2

3. Ask the child to point to all of the rectangular objects in the room.
4. Ask the child to draw a series of three geometric figures from memory. For example, say, "Draw a circle, a square, and a triangle."

5. Tell the child to circle the figure that is most like a given figure; for example, given ◯ , select from 𝟪 ఴ ౿ .

6. Tell the child to write what you dictate.
   a. three; tree
   b. nine; none
   c. seven; eleven

7. Ask the child to repeat a series of digits.
   a. 3 4
   b. 4 3
   c. 5 7 2
   d. 6 3 1 9
   e. 7 5 8 3 4

Teaching Activities

See Appendix C, pages 240–242, for the instructional strategies (IS) for Specific Learning Disabilities in Mathematics (SLDM) Checklist. This checklist may serve as a structure for developing an Individual Mathematics Plan (IEPM) for a student displaying one or more specific learning disabilities in mathematics. Following are some additional procedures for circumventing SLDM.

> ### Instructional Strategy 10.1.1:
> ### Use structured algorithms.

Some of the historical computational procedures help to circumvent errors due to misplacement of digits within an algorithm, reversed sequence of procedure, or messy writing. Reisman (1977) presented extensive examples of historical algorithms. Some of these are used here.[2]

Accommodate Reversed Sequence of Procedure

1. Retrograde method for addition (Reisman, 1977, p. 166):

> The Hindus used . . . the "inverse" or "retrograde" method. This involved adding from left to right, recording the results for each column, and blotting out or cancelling to show "carrying" from one column to the next, as shown below:

(a)	(b)	(c)	(d)
3	5	4	1
1	3	2	9
	3	1	2
4̸			
5	1	7̸	
		8	2

Step 1: 3 + 1 = 4. 4 is recorded in column *a*.

Step 2: 5 + 3 + 3 = 11. 1 is recorded in column *b*. 4 is crossed out and 4 + 1, or 5, is recorded in column *a*.

Step 3: 4 + 2 + 1 = 7. 7 is recorded in column *c*.

Step 4: 1 + 9 + 2 = 12. 2 is recorded in column *d*. 7 is crossed out and 7 + 1, or 8, is written in column *c*. Sum = 5,182.

2. Network method for multiplication (Reisman, 1977, p. 213):

> The "network" method of multiplication was very popular in the East and was adopted by the Arabs, who termed it *shabacah*, meaning "network," because of its reticulated appearance. (This process is the underlying principle of multiplying using Napiers rods.) The illustration below [Figure 10.1] shows the network method of multiplication.

> One factor (135) is written across the top of the grid and the other factor (12) is written vertically along the right side of the grid. Partial products are placed in the corresponding cells and are then summed along each diagonal.

> *Example:* 12 × 135

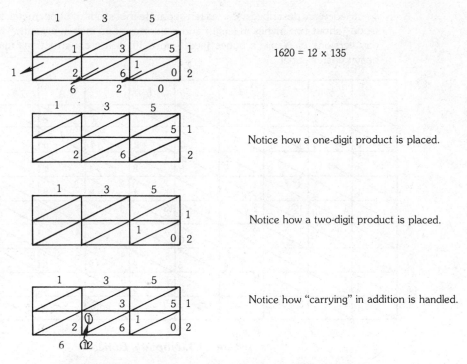

$1620 = 12 \times 135$

Notice how a one-digit product is placed.

Notice how a two-digit product is placed.

Notice how "carrying" in addition is handled.

Figure 10.1. Network Method of Multiplication

3. Quadrilatero: by the square (Reisman, 1977, p. 214):

The factors are placed one under the other as shown in the example below [Figure 10.2]. The multiplication then commences from left to right using as a multiplier 2 (2 × 5, 2 × 4, 2 × 3), then 7, and finally 9. The partial products are placed one under the other beginning each product in the ones position, recording from right to left. Then the diagonals are summed to obtain the answer. This method does not require attention to the places of the digits when performing the multiplication and may, therefore, aid a child who is having difficulty with place value in writing the modern multiplication algorithm.

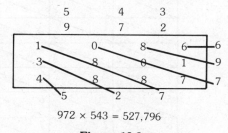

$972 \times 543 = 527,796$

Figure 10.2

4. Napier's Bones (Reisman, 1977, pp. 215-216):

In 1617, a Scotsman named Napier invented a calculating machine that was based on the network method of multiplication. "Napier's bones," as they have come to

179

be called, were described . . . as ten separate thin strips of bone, metal, ivory, or wood "about two inches in length and a quarter of an inch in breadth." Modern-day versions of Napier's bones [see Figure 10.3] are constructions made of a heavy poster paper.

1	2	3	4	5	6	7	8	9	R
1	2	3	4	5	6	7	8	9	1
2	4	6	8	10	12	14	16	18	2
3	6	9	12	15	18	21	24	27	3
4	8	12	16	20	24	28	32	36	4
5	10	15	20	25	30	35	40	45	5
6	12	18	24	30	36	42	48	54	6
7	14	21	28	35	42	49	56	63	7
8	16	24	32	40	48	56	64	72	8
9	18	27	36	45	54	63	72	81	9

Figure 10.3. *Napier's Bones*

In using Napier's bones to solve the problems 5 × 738 and 15 × 738, the 7, 3, and 8 rods would be placed side by side next to the guiding rod (R) [see Figure 10.4]. The product is found by adding in a downward movement within a diagonal in the section of squares that are in a row with the multiplier in the R column.

7	3	8	R
7	3	8	1
14	6	16	2
21	9	24	3
28	12	32	4
35	15	40	5
42	18	48	6
49	21	56	7
56	24	64	8
63	27	72	9

5 x 738 = 3690

15 x 738 = 11,070

Figure 10.4

Instructional Strategy 10.1.2:
Use hand calculator.

For those students displaying expressive disabilities, such as inability to say or write an answer on demand, use of a hand calculator allows circumvention of the problem. This is also a good drill activity—every time the child punches the keys and thinks "three plus five equals eight," he is practicing this basic fact.

Instructional Strategy 10.1.3:
Restrict number of computations.

Some children with specific learning disabilities tend to expend a great deal of energy when engaged in fine motor activities—for example, writing computations. Such children may produce an excess of saliva and/or show a noticeable degree of fatigue half-way through such activities. Therefore, give them only enough computations to permit them to demonstrate their knowledge. Three to five computations within an operation should be sufficient and should control for guessing or random errors. A page of 20 or more computations is punishing.

Instructional Strategy 10.1.4:
Develop a time schedule for completing a task.

Some children with specific learning disabilities seem to be in a constant battle with time. They often wait until the last minute to study for an exam; they attempt to write a term paper the night before it is due; or they are unable to pace themselves during a task, taking too much time on some items and not enough on others—or else they don't finish at all. The teacher can help such children write a time schedule and can remind them when time sequences are up. A fading-out technique, utilizing the child's own attention to the end of a predetermined sequence for a portion of a task, builds independent pacing by the child. By teaming tasks and tests, a peer can be used to help such a child with his pacing schedule.

Instructional Strategy 10.1.5:
Use tactile experiences.

The child's ability to use symbols—the digits, the operation signs, the equality and inequality signs and the set operations signs, all of which are arbitrary associations—is affected by visual perception disabilities. To compute basic facts, a child must have a working knowledge of these symbols. Some children need practice with tactile experiences (tracing clay or cut-out symbols) in order to produce these symbols correctly.

181

> ### *Instructional Strategy 10.1.6:*
> ### *Use fixed representation of number.*

Using fixed examples of a number—dominoes, playing cards, or pictures of objects—helps to circumvent visual perception disabilities. The physical appearance or spatial arrangement of objects used can make the number property of equivalent sets hard to detect.

Summary Thought Questions and Activities

I.   Tell why the algorithms in IS 10.1.1 are appropriate for obtaining the correct results. Also, compare the historical algorithms presented with presently used algorithms found in any children's mathematics text. Consider number of steps needed to process the numbers, likelihood of error, and simplicity/complexity of figure–ground aspects.

II. Using the SLDM checklist (Appendix C, pages 240-242), assess three primary or upper grade children after observing each for a period of time as they were engaged in various activities. If possible, compare your assessments with that of the child's teacher.

III. List the behaviors of the children observed in Activity II. Develop a set of inferences that may serve as explanation for these behaviors. Relate the instructional strategies suggested in this chapter to these inferences; this should make evident the rationale for each IS.

Suggested Readings

Bannatyne, A. *Language, reading and learning disabilities.* Springfield, Ill.: Charles C. Thomas, 1971.

Cruickshank, W. M. *Psychology of exceptional children and youth* (3rd ed.). Englewood Cliffs, N.J.: Prentice-Hall, 1971.

Frostig, M., & Maslow, P. *Learning problems in the classroom.* New York: Grune & Stratton, 1973.

Ginsburg, H. *Children's arithmetic.* New York: D. Van Nostrand Co., 1977.

Smith, I. M. *Spatial ability: Its educational and social significance.* London: University of London Press, 1964.

References

Reisman, F. K. *Diagnostic teaching of elementary school mathematics: Method and content.* Skokie, Ill.: Rand McNally, 1977.

Reisman, F. K., Braggio, John T., Braggio, S. M., Farr, J. & Simpson, L. *Physiological measures of arithmetic performance in children.* Mathematics Information Report, The ERIC Science, Mathematics and Environmental Education Clearinghouse Center for Science and Mathematics, Ohio State University, April 1977.

Endnotes

1. Reisman et al. (1977) tested children in elementary school grades who had been referred to their school's learning disabilities program. They observed physiological measures of laryngeal activity and two measures of respiration (amplitude and frequency) while children performed computations involving adding, multiplying, subtracting, and dividing. Low-performing children on the mathematics tasks displayed generally high levels of activity on physiological measures, regardless of the difficulty of the items. On the other hand, children who obtained the highest scores on the computation test showed a pattern of physiological activity that increased and decreased in relationship to the difficulty of the item.

2. From *Diagnostic teaching of elementary school mathematics: Method and content* by F. K. Reisman, © 1977 Rand McNally. Used with permission.

Instructional Strategies Based on Sensory and Physical Needs

11 In keeping with the function approach, as opposed to the approach of labeling or categorizing exceptionalities, the instructional alternatives presented allow the teacher to adapt mathematics instruction to the student's intact modalities. The emphasis is on circumventing handicaps rather than on focusing on effects of the handicap. It should be noted, however, that in this chapter an average rate of cognitive development is assumed. For those students whose special educational needs are compounded by multiple handicaps—for example, blind-retarded—the teacher is expected to select instructional alternatives appropriate to the compounded condition. (The teacher should refer to the Diagnostic Guide, pages 296–309, which cross-references each instructional strategy by related generic factors and relevant mathematics topics.)

11.1.0 Impaired Hearing

Hearing-impaired students have missed out on information that is ordinarily acquired auditorily. Thus, it is important to assess whether the student has the experience requisite for understanding what is to be learned. It may be necessary to simplify statements and questions, taking care not to cause misconceptions that often result when textual material is simplified. Language acquisition problems also influence development of mathematics as a language of relationships. Language errors indicative of a hearing impairment involve meanings of words, meanings of idioms, meanings due to intonations, questions, syntax, figurative language, making comparisons and inferences, and drawing conclusions.

Some general suggestions for teaching the hearing-impaired include the following:

Use pictures, diagrams, or objects to concretize instruction.

Make sure mathematics vocabulary is systematically taught by providing explanations and examples.

Relate indefinite time relationships (long ago, in a little while, next time) to concrete examples.

Point out differences in such words as one and won; *two and* to *and* too; *three and* tree; *four and* for; *eight and* ate; *seven and eleven.*

Point out inflectional endings, such as plurals and tense markers.

Assessment Activities: Chapter 2, sections 2.2.0 and 2.3.0 present examples that may be used as assessment activities for the hearing-impaired.

Teaching Activities

The instructional strategies for hearing-impaired students emphasize that mathematics is a language involving relationships, concepts, generalizations, and their symbolic representation. Therefore, there is more at stake for the hearing-impaired student than learning to solve word problems. The entire developmental mathematics curriculum is related to the student's language acquisition. Realizing that a spoken word or a graphic symbol represents an idea is a breakthrough for the hearing-impaired, especially for those who have had the handicap from infancy. Such an awareness underlies the entire notion of symbolic representation.

> ### Instructional Strategy 11.1.1:
> #### Use cues.

The use of visual aids such as color or increased size for learning arbitrary symbols helps to circumvent hearing impairment. See also the following instructional strategies: 5.2.1, 5.2.2, 5.2.3, 5.3.2, 5.4.1, 5.4.2.

> ### Instructional Strategy 11.1.2:
> #### Use Fitzgerald Key.

The Fitzgerald Key (1954) is a chart of symbols that serve as visual cues for developing sentence structure, word order, word meaning and word modifiers. This strategy is particularly appropriate for teaching word problems. Commonly used symbols are as follows:

Noun	Pronoun	Verb	Conjunction	Infinitive	Present Participle

Prior to using the Key, students classify words they have learned (through vocabulary lessons) according to the following: *who, what, how many, where, adjectives* and *verbs* (grouped according to the Fitzgerald Key symbols).

185

Following are examples of the Fitzgerald Key applied to mathematics instruction. Have students classify mathematics vocabulary by the Fitzgerald Key symbols as follows:

noun	verb	adjective	conjunction
—	=	⌐¬	⌐↗
"I have *one*"	"*add* four"	"*plus* sign" "*one* dress"	"three *and* two"

who	verb	how many	what
⊤	=	⌐¬	—
She	has	two	dogs

> ## *Instructional Strategy 11.1.3:*
> ### *Use logographs.*

Logographs are symbols that stand for real objects or ideas. (Review pages 7–9.) Logographs may express quantitative and spatial ideas.

> ## *Instructional Strategy 11.1.4:*
> ### *Use classroom management techniques.*

The following general class organization techniques should be used when teaching the hearing-impaired:

> *Provide opportunities for small-group work to insure that the hearing-impaired child actively communicates.*

> *Seat the child so she can see the teacher most of the time. Have the light (windows) in back of the child to facilitate unobstructed use of visual cues.*

> *Seat the child so that most class noise is behind her. This minimizes interference from extraneous noise which can be very annoying, especially for those wearing a hearing aid.*

> *Get the child's attention first by moving close. Then speak slowly, clearly, and distinctly, but don't shout at the child.*

> *Rephrase instruction or ideas until you are sure they were understood.*

> *Provide for maximum use of the child's residual hearing.*

> *Assign a buddy to help the hearing-impaired child by interpreting directions, questions, and assignments, and taking notes.*

> *Provide means for previews or preteaching prior to class lessons via buddy or resource teacher. For example, before a lesson, write new mathematics vocabulary for the child and make sure each new word is understood. Give an outline of the lesson to the hearing-impaired child to facilitate his following the presentation.*

> ### *Instructional Strategy 11.1.5:*
> ### *Use multisensory approaches.*

Utilize all senses by providing verbal, auditory, visual, and tactile experiences. This approach may be applied to such topics as money, time, weights and measures, calendar, geometric figures, reading/writing digits and numerals, operations with fractions, percentage graphs, scales.

> ### *Instructional Strategy 11.1.6:*
> ### *Use media.*

The following suggestions for using media to implement a multisensory approach were presented by Nober, Alves, Clark, Helfgott, Hunt, Petrie, and Weiner (1975):

Use special boards that provide visual or tactile stimulation: clean chalkboards, bulletin boards, flannel/magnetic boards.

Use self-instructional devices: filmstrip viewers, cartridge projectors, programmed learning machines.

Use a variety of printed materials: supplementary books, reference books, periodicals, newspapers, and programmed materials.

Use materials especially designed for visual or tactile learning: charts, maps, globes, photographs, games, specimens, models, diagrams.

Use the overhead projector. It has two advantages. The teacher faces the students rather than facing the blackboard, and the room need not be darkened. The light that is cast on the teacher's face facilitates speech reading.

Use captioned filmstrips and slides. To make sure that the hearing-impaired child has read each caption, the teacher may instruct the student to change each picture after reading the caption.

If a cassette tape or record is used, supply the hearing-impaired child with a script.

During a filmstrip or slide presentation, a flashlight should illuminate the teacher's mouth if he speaks. Extra time should be allowed for the child to both speech read and view the picture.

Use a pointer to direct attention to the portion of an equation or geometric figure being discussed.

A topic outline, summary, or script should be given the hearing-impaired child in advance of viewing a film.

When using a tape recorder or record player for auditory training, students should use headphones or appropriate amplification to minimize extraneous sounds. When a group is listening to a recording for informational purposes, the hearing-impaired child should be engaged in an alternative activity to meet the same objective.

> ## *Instructional Strategy 11.1.7:*
> ### *Use familiarity.*

As mentioned earlier, these children have difficulty in dealing with semantic aspects of language, particularly word meanings and figurative language. It is necessary to build concepts on familiar ideas. This instructional strategy is closely related to instructional strategies 5.2.2 and 11.1.5; these strategies emphasize familiarity with real-life experiences that give meaning to mathematical relationships and concepts. Indefinite time relationships such as "in a little while" or "I'll be there in a second" or "next time" either are taken literally or have no meaning at all without some familiar experience to support them. For example, to give some idea of infinity, have the child view the horizon or look at the sky. To represent the null set, have him empty a cellophane bag of candy and retain the empty bag.

11.2.0 Impaired Vision

Listening and memory skills must be sharpened for the visually impaired learner, and tactile experiences are important. The current trend is to divide the visually impaired into those who can read large print (some need the aid of a variety of devices, such as magnifiers or light sources) and those who can read only Braille. Many children classified as legally blind have been found to be able, with training, to use residual sight to read large-print materials.

Suggestions for instructional strategies of a general nature are as follows:

A child with low vision may have to approach the chalkboard to see an equation or the like. She may have to hold written material very close to the face when reading.

A visually impaired child does not have a memory of how things look and therefore will not make use of visual imagery, but must rely on tactile and auditory perceptions.

Provide much verbal explanation; it is not necessary, though, to speak more loudly to a blind person than to anyone else.

Provide experiences with nonvisual cues that allow senses other than vision to be used for noticing topological ideas, such as proximity and order. Use of a tactile map is an example.

Assessment Activities: These are the same as for Chapters 5 through 9, with modifications in material as suggested in the instructional strategies that follow.

Teaching Activities

> **Instructional Strategy 11.2.1:**
> *Use vision aids.*

Vision aids include magnifiers, monocular lenses, and large-print materials.

> **Instructional Strategy 11.2.2:**
> *Use nonvisual cues.*

Touch: Have the vision-impaired child trace around shapes, fold paper into different shapes and sizes, and select objects that are described by their shape.

Sound: Establish directionality by having the vision-impaired child point to the location of another child talking or making a specified sound such as ringing a bell, stamping foot, dropping a book. Child may also *count* the number of sounds he hears.

Smell: Proximity can be identified by having the child tell if something is nearby (flowers, freshly baked bread), or if some kind of work is occurring nearby (tarring road, painting, freshly mowed lawn).

Taste: Give the child small amounts of orange juice that progress in taste from very much watered down to straight frozen concentrate. Point out the different intensities of taste. Of course, be sure that the student is not allergic to the orange juice (chocolate, salted food) that you ask him to taste. This activity also stresses similarities and differences.

> **Instructional Strategy 11.2.3:**
> *Use Braille numerals.*

Six dots form the structure of the Braille code. By varying the position and number of embossed dots within each three-by-two array, the letters of the alphabet and the numbers 1 through 10 may be represented. The mirror image for the Braille symbol for the letter v, placed before a character, indicates that it represents a digit and not a letter (the letters a through j thus stand for the numbers 1-10). The Braille code is shown in Figure 11.1.

11.3.0 Physical Impairment

As mentioned earlier in Chapter 3, section 3.1.0, many mathematical experiences are missing for children with a physical impairment. Restriction of

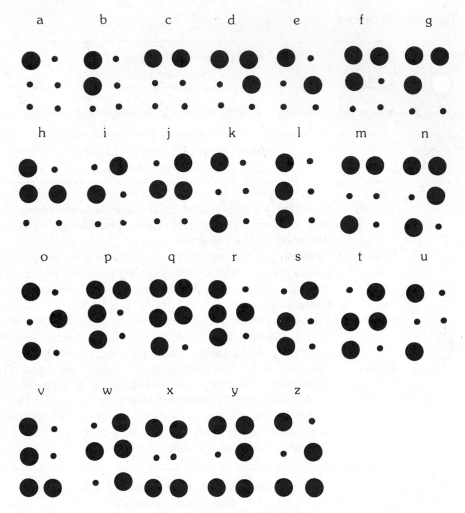

Figure 11.1. *Braille Code*

Adapted from the Visual Impairment Unit published by SEE.

activity that accompanies physical impairment may also limit opportunities to develop spatial and temporal relationships in particular.

Assessment Activities:

Mathematics ideas presented in Chapters 5 through 9 that are especially important as a foundation for mathematics, but are often not experienced by those with a physical impairment, include: money, measurement, size, shape, and mathematical language (big–little, longer–shorter, heaviest–lightest, prepositions). Assessment activities discussed in Chapters 5-9 for these topics may be modified by emphasizing verbal expression or gestures in order to circumvent a physical disability.

Teaching Activities

A child who has been immobile from infancy because of a physical impairment has not experienced the movement in space necessary for developing spatial and temporal relationships and concepts. The rhythm involved in crawling and walking is mathematical, in a temporal sense. These activities also involve succession and sequencing as the infant moves his hands and feet to crawl. The teacher must encourage and point out movements in space—even movements in a wheelchair—in order to provide the prerequisite experiences for directionality, enclosure, proximity, and order. The *temporal* aspect of number must also be considered. The teacher for the physically handicapped can provide instructional materials to help circumvent various physical problems. For example, the child who has no limbs can point with a head pointer; there are stands to hold textbooks, mechanical page turners, templates, line guides, standard or modified typewriters, and hand calculators.

(IS)

> ### Instructional Strategy 11.3.1:
> *Minimize manual dexterity in performance requirements.*

Use tasks that involve verbal or gestural performance, such as computations with hand calculators or on a typewriter, if it is difficult for a child to write legibly.

(IS)

> ### Instructional Strategy 11.3.2:
> *Adjust the pace of instruction.*

Children with a physical impairment may become fatigued more quickly and should be encouraged to set their own time limits for task completion. Those with problems of mobility, slowness in graphic skills, or orthopedic problems may need more time to progress through a learning experience.

(IS)

> ### Instructional Strategies 11.3.3:
> *Use team taking of tests.*

Teaming a peer with a child who cannot write legibly due to a physical handicap creates a situation in which each contributes to learning during test taking. The team testing procedure has been used by one of the authors in classroom situations in grades two, three, five, high school geometry, and undergraduate and graduate mathematics content and methods courses. Attitude measures (pre- and post-test) showed a decrease in test anxiety and lessened mathematics anxiety. Team members reported that they studied harder so as not to "let their partner down" or "look like a dummy" to their partner. Procedures for instituting teamtesting are as follows:

1. *Allow children to team with a partner.*

2. *Explain that they should feel free to change partners for the next test if they believe their partnership was not bringing out their best performance. Also, they can work alone if preferred.*

3. *They may handle the test in whatever way is most comfortable. For example, some go through an entire test independently and then compare answers, while other teams discuss each item as they go.*

4. *They may disagree on answers and indicate by initialing each one's answer for those items so the teacher knows who belongs to which response.*

11.4.0 Low-Vitality and Fatigue Conditions

Effects of low-vitality conditions on learning mathematics should be reviewed at this time (see Chapter 3). It should be remembered that some children may tire from *thinking.* Much of mathematics involves abstract thinking. A child gifted in mathematics, if he or she has a medical problem that causes rapid fatigue, may become exhausted after a short time even though engaged in a task that is enjoyable.

The assessment and teaching activities presented in Chapters 5 through 9 may be used for a student with a low-vitality or fatigue condition.

Summary Thought Questions and Activities

I. Become acquainted with materials for teachers and parents of the hearing-impaired, including: *The Volta Review, Exceptional Parents Magazine, The American Annals of the Deaf,* and *Ideas for Parents* (a newsletter published by the Lexington School for the Deaf, 30th Avenue at 75th Street, Jackson Heights, New York 11370).

II. Use excerpts from Hamilton, Howard, Bornstein, Miller, and Sauliner (1975) to practice signing. (See Figures 11.2–11.4.)

III. Perform computations using the Braille numerals as shown in Figure 11.1.

Suggested Readings

Barraga, N. *Visual handicaps and learning: A developmental approach.* Belmont, Calif.: Wadsworth, 1976.

Bitter, G. B. (Ed.) *Systems O.N.E.* Salt Lake City: University of Utah Press, 1974.

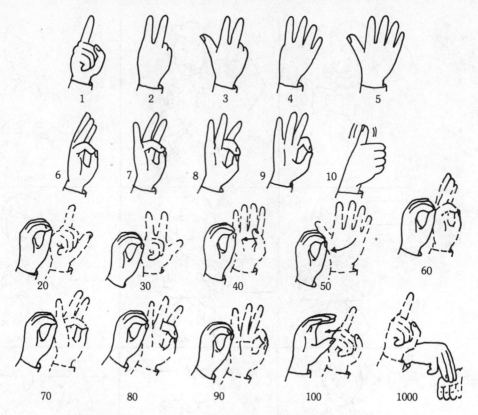

Figure 11.2.   *Numbers*

Figures 11.2, 11.3, and 11.4 are adapted from *Signed English in the classroom* by L. B. Hamilton, R. L. Howard, H. Bornstein, R. R. Miller, & K. L. Sauliner, © 1975 by Gallaudet College Press. Used with permission.

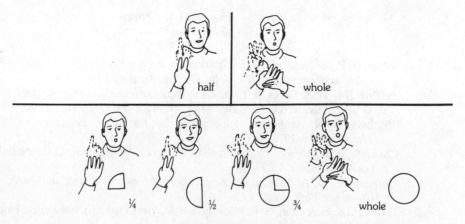

Figure 11.3.   *Fractions*

Source: See Figure 11.2.

circle   diamond   heart

rectangle   square   triangle

curve   round   star

Figure 11.4. Shape
Source: See Figure 11.2.

Bitter, G. B., & Mears, E. G. Facilitating the integration of hearing impaired into regular public school classes. *The Volta Review*, January 1973, 75, 13–22.

Boatner, M. T., & Gates, J. E. (Eds.) *A dictionary of idioms for the deaf.* Washington, D.C.: Alexander Graham Bell Association for the Deaf, Inc., 1966.

Brothers, R. J. Arithmetic computation by the blind. *Education of the Visually Handicapped*, 1972, 4, 1–8.

Caravella, J. R. *Minicalculators in the classroom.* Washington D.C.: National Education Association, 1977.

Christopher, D. A. *Manual communication: A basic text and workbook.* Baltimore: University Park Press, 1976.

Davis, H., & Silverman, S. R. *Hearing and deafness* (3rd ed.), New York: Holt, Rinehart and Winston, 1970.

Furth, H. G. *Deafness and learning: A psycho-social approach.* Belmont, Calif.: Wadsworth Publishing Co., 1973.

Hanninen, K. A. *Teaching the visually handicapped.* Columbus, Ohio: Charles E. Merrill Company, 1975.

Lowenfeld, B. (Ed.). *The visually handicapped in school.* New York: Day, 1973.

McDonnell, F., & Ward, P. (Eds.). *Deafness in childhood.* Nashville, Tenn.: Vanderbilt University Press, 1967.

Myklebust, H. R. *The psychology of deafness: Sensory deprivation, learning and adjustment.* New York: Grune and Stratton, 1964.

Napier, G., Kappan, D. L., Tuttle, D. W., Schrotberger, W. L., & Dennison, A. L. *Handbook for teachers of the visually handicapped.* Louisville: American Printing House for the Blind, 1975.

Northcott, W. H. *The hearing impaired child in the regular classroom.* Washington, D.C.: Alexander Graham Bell Association for the Deaf, Inc., 1973.

Reynolds, M. C., & Birch, J. W. *Teaching exceptional children in all America's schools.* Reston, Virginia: The Council for Exceptional Children, 1977.

Scholl, G. T. The education of children with visual impairments. In W. M. Cruikshank & G. O. Johnson (Eds.), *Education of exceptional children.* Englewood Cliffs, N.J.: Prentice-Hall, 1975.

Watson, T. J. *Education of hearing handicapped children.* Springfield, Ill.: Charles C. Thomas Publisher, 1967.

Watson, D. (Ed.) *Readings on deafness.* New York: New York University, Deafness Research and Training Center, 1973.

References

Fitzgerald, E. *Straight language for the deaf: System of instruction for deaf children* (2nd ed.). Washington, D.C.: The Volta Bureau, 1954.

Hamilton, L. B., Howard, R. L., Bornstein, H., Miller, R. R., & Sauliner, K. L. *Signed English for the classroom.* Washington, D.C.: Gallaudet College Press, 1975.

Nober, L., Alves, S., Clark, A., Helfgott, S., Hunt, D., Petrie, S., & Weiner, S. *Hi-Fi.* Washington, D.C.: National Audiovisual Center, 1975.

195

Instructional Strategies Based on Social and Emotional Needs

12 Social and emotional factors that may affect mathematics learning are considered in the following questions.

- What is the degree of independence that the child displays in various situations?
- How is the child handling his or her handicap?
- Is the child presented mathematics goals that are achievable in terms of his limitation?
- Are the child's teachers realistic in selecting curriculum that is not too difficult but also not so easy that tasks are insulting to the student?
- Are realistic mathematics goals communicated to the child's parents so that the parents may reinforce classroom instruction?
- Are teachers consistent in managing the child's behavior?
- Is instruction offered in relevant skills?
- Are contingencies of reinforcement appropriate?
- Are school administrators, custodians, and other students sensitive to the child's needs?

In order to break the "mathematics—ugh!" cycle that pervades responses of people of all ages to mathematics, teachers must develop an awareness of their own role in causing these responses. Of course, parents must share in this responsibility by becoming aware of ways in which their own attitudes and behavior compound their youngster's fear and discomfort in mathematics activities. It is amazing how often an adult who expresses fears and insecurities concerning ability in mathematics nevertheless is insensitive to these same feelings in his own children or students.

In Chapter 4, clusters of disordered behavior presented by Kauffman (1977) were listed. They form the structure of this chapter; general instructional strategies are grouped according to these four clusters of generic factors that are emotional and social in nature.

12.1.0 Hyperactivity, Distractibility, and Impulsivity

Review Chapter 4, section 4.1.0.

> **Instructional Strategy 12.1.1:**
> **Use behavior modification techniques.**

Reward the child's desirable behavior and follow his or her undesirable behavior with either no consequences or punishing consequences. This may be as simple as shifting teacher attention from inappropriate to appropriate behavior. More complicated techniques involve token reinforcement or earning points for desirable behavior. These tokens or points may be exchanged for reinforcers such as candy, school supplies, free time, or for whatever the child agrees would be worth changing his behavior for.

> **Instructional Strategy 12.1.2:**
> **Provide a structured environment.**

A highly structured classroom has the following characteristics (Cruickshank, Bentzen, Ratzeburg, & Tannhauser, 1961; Haring & Phillips, 1962; Haring & Whelan, 1965; Hewett, 1968; Kauffman, 1977):

- A minimum of distracting stimuli are in the classroom.
- Clear expectations are developed with regard to the children's movements about, and classroom routine is predictable.
- Consistent consequences are applied to hyperactive and nonhyperactive behavior.
- The teacher is highly directive making most decisions until the child can manage himself.
- The mathematics environment also must be structured by controlling use of manipulative materials. For students who do not exert self-control, manipulatives should be used in a one-to-one or peer-team situation lest they end up being thrown around the room, only having tempted the child into trouble. The purpose of materials and rules for their use should be made clear to the child.

> **Instructional Strategy 12.1.3:**
> **Use self-instruction.**

Children can learn to control their excessive motor behavior by giving themselves instruction (Goodwin & Mahoney, 1975; Hallahan & Kauffman, 1976; Luria, 1973). The student who is easily distracted can be instructed to talk himself through a mathematics task. Those who have difficulty staying on task while doing computations can be prevented from sidetracking by self-monitoring.

> **Instructional Strategy 12.1.4:**
> **Eliminate irrelevant or potentially distracting stimuli from instructional materials.**

Eliminate distracting colorful pictures or diagrams in mathematics textbooks by cutting out what is salient and pasting it on a plain sheet of paper. Present arithmetic computations one at a time. Use cues such as color or texture to highlight relevant and salient parts of a problem—for example, for the child who needs help in noticing if the operation sign is + or -. At the same time, eliminate irrelevant stimuli that will be hard to ignore.

> *Instructional Strategy 12.1.5:*
> *Use modeling.*

Children learn by imitating adults or peers in their environment. Undesirable behavior—hitting, kicking, poking—can be reduced in elementary school children by inducing attention to, and reinforcing, desirable behavior (Csapo, 1972; Cullinan, Kauffman, & LaFleur, 1975; Goodwin & Mahoney, 1975).

Adults or peers who consistently point out consequences or implications of situations help the behavior-disordered student develop sequencing of thought. Developing graphs that record probable outcomes provides an activity that interests and motivates middle school and high school students in particular. For the hyperactive child, graphing instances of desirable behavior in comparison with undesirable helps the student obtain an objective view of his or her behavior. Such an activity may be modeled by peers whom the student admires or is friends with.

12.2.0 Aggressiveness

Review Chapter 4, section 4.2.0.

> *Instructional Strategy 12.2.1:*
> *Select successful mathematics task to obtain high success rates.*

The student who has episodes of aggression also has quiet times when he or she is willing to engage in an activity that provides a measure of success. It is at these times that the teacher must select mathematics activities that ensure success for the student. The Screening Checklist, Appendix D, can provide a general idea of the student's strengths and weaknesses in mathematics performance. Use this checklist, along with task analysis of a goal selected for the student, to identify a relevant topic.

12.3.0 Withdrawal, Immaturity and Inadequacy

Instructional strategies for these children would be the same as some of those described previously in this chapter, such as modeling and behavior modification. The main focus is to teach the child who is socially isolated to reciprocate positive social responses—to teach the child how to become involved with peers and adults.

> ### Instructional Strategy 12.3.1:
> #### Arrange appropriate environmental conditions.

Provide equipment or toys conducive to social play, and bring the isolate in proximity to peers who have good social interaction skills. A store in the room, where children can refine their use of money, can help a withdrawn child. Team-taking of mathematics tests is also suggested as a socializing activity.

> ### Instructional Strategy 12.3.2:
> #### Teach recognition of important social cues.

Showing the child videotaped replays of his own behavior and then role playing and modeling appropriate behavior help to teach the child to recognize such cues as when to be quiet, when to listen, when to take one's turn, when to cooperate. You may have children role-play (a) buying clothes in a store, (b) interacting with a plant foreman who is showing how a time clock works, or (c) having a husband-wife discussion regarding budgeting, saving, banking, and so on.

     The mathematics may serve two roles. First, it may be the vehicle for dealing with withdrawal, immaturity or feelings of inadequacy; in this case, incidental learning may or may not occur. In this circumstance, the instruction is focused on reducing the undesirable behavior. On the other hand, the child may become really interested in the role-play situation, and instruction may then be focused on the pros and cons of buying on credit, budgeting, etc.

> ### Instructional Strategy 12.3.3:
> #### Insure success in tasks.

"Disturbed children refuse even to try—they are defeated in their own minds before they give themselves any opportunities for success" (Kauffman, 1977, p. 239). The use of task analysis of mathematics goals enables the teacher to identify components of a task. Then the teacher can present a step leading to a goal that the child already has in his or her repertoire, insuring that the child will be successful. This shaping of the self-concept encourages the child to try, to interact.

12.4.0 Deficiencies in Moral Development

Review Chapter 4, section 4.4.0.

> ### *Instructional Strategy 12.4.1:*
> *Sensitize students to moral issues.*

The teacher may enlist the cooperation of parents and peers to use discipline, modeling, reasoning, and encouraging the student to take another's role in situations of moral conflict. Topics such as economics, statistics, graphs and measurement are appropriate as vehicles for sensitizing students to moral issues. For example, the issues of drug use, establishing credit, or keeping up interest payments may be relevant.

Summary Thought Questions and Suggested Activities

I.   Assess a child who is displaying behavior disorders according to the questions presented on page 196.
   A.   Describe those situations in which the child displays independence.
   B.   Describe your observations concerning how the child handles handicapping conditions.
   C.   Identify mathematics goals that do not seem realistic for the child and explain why you believe this.
   D.   Select one of the Instructional Strategies in Chapter 12 and try it with this child.

II. Visit a school classroom and evaluate the structure of its environment during mathematics activities. Use the suggestions in IS 12.1.2 as a guide.

III. Identify one of your own mathematics teachers who has served as your model. List those characteristics that you find yourself emulating.

Suggested Readings

Bandura, A. *Aggression: A social learning analysis.* Englewood Cliffs, N.J.: Prentice-Hall, 1973.

Clements, J. E., & Platt, J. The yes-no graph: A parent involvement tool. *Academic Therapy*, 1978, 323–328.

Ebmier, H., & Good, T. L. The effects of instructing teachers about good teaching on the mathematics achievement of fourth grade students. *American Educational Research Journal*, 1979, *16*, 1–16.

Good, T. L., & Brophy, J. E. Changing teacher and student behavior: An empirical investigation. *Journal of Educational Psychology*, 1974, *65*, 390–405.

Heady, J., & Niewoehner, M. Academic and behavior management techniques that work. *Teaching Exceptional Children*, 1979, *12*, 37–39.

Kohlberg, L. Development of moral character and moral ideology. In M. L. Hoffman & L. W. Hoffman (Eds.), *Review of child development research* (Vol. 1). New York: Russell Sage Foundation, 1964.

Kohlberg, L. Education for justice: A modern statement of the Platonic view. In N. F. Sizer & T. R. Sizer (Eds.), *Moral education: Five lectures.* Cambridge, Mass.: Harvard University Press, 1970.

Kohlberg, L. Stage and sequence: The cognitive-development approach to socialization. In D. Goslin (Ed.), *Handbook of socialization theory and research.* Chicago: Rand McNally, 1969.

Markman, E. M. Realizing that you don't understand: A preliminary investigation. *Child Development,* 1977, 47, 986–992.

Schreibman, L. Effects of within-stimulus and extra-stimulus prompting on discrimination learning in autistic children. *Journal of Applied Behavior Analysis,* 1975, 8, 91–112.

Wood, M. M. *Developmental therapy.* Baltimore: University Park Press, 1975.

References

Cruickshank, W. M., Bentzen, F., Ratzeburg, F., & Tannhauser, M. A. *A teaching method for brain-injured and hyperactive children.* Syracuse, N.Y.: Syracuse University Press, 1961.

Csapo, M. Peer models reverse the "one bad apple spoils the barrel" theory. *Teaching Exceptional Children,* 1972, 5 (1), 20–24.

Cullinan, D. A., Kauffman, J. M., & LaFleur, N. K. Modeling: Research with implications for special education. *Journal of Special Education,* 1975, 9, 209–221.

Goodwin, S. E., & Mahoney, M. J. Modification of aggression through modeling: An experimental probe. *Journal of Behavior Therapy and Experimental Psychiatry,* 1975, 6, 200-202.

Hallahan, D. P., & Kauffman, J. M. Research on the education of distractible and hyperactive children. In W. M. Cruickshank & D. P. Hallahan (Eds.), *Perceptual and learning disabilities in children* (Vol. 2: *Research and theory).* Syracuse, N.Y.: Syracuse University Press, 1975.

Haring, N. G., & Phillips, E. L. *Educating emotionally disturbed children.* New York: McGraw-Hill, 1962.

Haring, N. G., & Whelan, R. J. Experimental methods in education and management. In N. J. Long, W. C. Morse, & R. G. Newman (Eds.), *Conflict in the classroom.* Belmont, Calif.: Wadsworth, 1965.

Hewett, F. M. *The emotionally disturbed child in the classroom.* Boston: Allyn & Bacon, 1968.

Kauffman, J. M. *Characteristics of children's behavior disorders.* Columbus, Ohio: Charles E. Merrill, 1977.

Luria, A. R. *The working brain: An introduction to neuropsychology.* New York: Basic Books, Inc., 1973.

Implementation

SECTION III

Four specific competencies have been identified that are necessary for mathematics teachers of the handicapped and the talented. These include the following:

1. *interviewing techniques*
2. *knowledge of Public Law 94-142: The Education for all Handicapped Children Act*
3. *skills in developing individual mathematics programs (IEPMs)*
4. *competence in working as a member of an interdisciplinary education team including parents.*

Information on Public Law 94-142, technical aspects of developing individual educational programs (IEPs), descriptions of IEP committee activities, and how teachers and parents can help one another for the benefit of a child are presented in the following section.

Interviewing Techniques

13 The interviewer's principal tool is himself, and as we all differ from one another, we are all using different tools. Slavish imitation of somebody else's technique is like buying a suit or dress because it looks nice on somebody else. (Rich, 1974, p. viii)

Functions of Interviews

Interviews often change their character—that is, their function—as they progress. Interview functions include the following: fact-finding, fact-giving, manipulative, and treatment. (Rich, 1974).

An inherent danger in an interview function that is primarily *fact-finding* is that the real problem may be missed.[1] This occurs especially with children because the roles of the adult interviewer and the child are unequal. The child may answer questions put forth but may not feel free enough to elaborate on points that the teacher did not think of asking. For example, the teacher may say, "Is there anyone who does not understand?" and the child may not know what it is she does not understand. Another danger in the fact-finding function of an interview is collecting a large number of facts that may be relevant but that represent an "overkill." Most people dislike having to answer the same question more than once.

The main distinction between *fact finding* and *fact-giving* interviews is who decides which facts are to be given. A teacher who is lecturing is engaged in a fact-giving interview. In this type of interview, the person receiving the information must *want* to receive—he or she must be tuned in. In teacher–child interviews, the adult must avoid preventing the child from fact-giving. Adults sometimes insist upon engaging in fact-finding of irrelevant data, to the point of turning off the young interviewee.

The child who fails a mathematics test and bursts out crying does not need a teacher to carry on a fact-finding interview at that moment. Instead, the teacher may encourage the child to move on to an activity in which he is successful—for example, helping to plan a student–faculty play. With regard to the mathematics test, the teacher may simply say, "we'll work this out together." The teacher thus is conducting a *manipulative* interview, manipulating the child into a certain frame of mind. Rich (1974, p. 11) has noted that "If [*the teacher*] had started looking for facts at this point, she would probably have made the child even . . . sorrier for himself, and [the teacher] would not have learned the facts anyway."

Supportive, counseling, and *insight-giving* interviews may be considered treatment interviews. In supportive interviews, the teacher listens, encourages, and sympathizes. The counseling interview is a rational discussion of clearly stated problems. Although evident to the interviewer, the student's unconscious motivations are not dealt with directly. Insight-giving interviews deal with unconscious motivations; the appropriate techniques are beyond the expertise of most school personnel.

Rich (1974, p. 42) suggested the following as a summary of guidelines for interviewing children:

> Most children do not want to talk about school, but do want to talk about their leisure activities and hobbies. The interview should therefore begin with these rather than school topics. The key to successful communication is having a common ground with the child.

Types of Questions

Questioning skills are an important part of interviewing technique. An interviewer may use several types of questions including questions that are direct, indirect, convergent, and divergent.

Direct questions assume that the child knows the answer to what he is being asked.

Examples
   Convergent questions require a simple, straight answer (such as yes or no):
   How old are you? Do you like your teacher? What is your favorite sport? Do you like arithmetic?
   Divergent questions require essay-type answers: What do you think of mathematics?

Indirect questions often provide information that a direct question might block. By asking two questions separated by several minutes, the information being sought is not so obvious. An example of this procedure is to ask, "What topics in arithmetic do you think are most important to learn?" and later on, "What topics in arithmetic are you learning?" A student may answer "how to use a calculator for shopping" or "fractions for building birdhouses to sell" to the first question and "long division" to the second; yet he may deny that he is uninterested in class if this point is put to him directly.

Indirect questions allow the teacher to obtain students' opinions concerning situations they may not address directly. For example, a fourth grade girl with cerebral palsy, whose fine motor skills made written computation difficult, refused to use a hand calculator placed on a table for her convenience. Indirect questioning elicited that she thought her teacher disapproved of using hand calculators. When this was clarified by her teacher, the young lady smiled and said, "Oh, I'm so glad, because it's really tiring when I have to do so much writing."

A *leading* question is one in which the answer is suggested: "You don't like arithmetic, do you?" This type of question is legitimate if it is clear what the student is trying to say and the interviewer is merely putting the student's thought into words. However, the interviewer must be careful not to put words in the child's mouth—making him say something he does not mean. A leading question may also be in the form, "Compared to other subjects, is arithmetic clear or unclear most of the time?" "Or multiple-choice questions may be posed: "Which word would best show how you feel about arithmetic: happy, sad, about even?"

Rich (1974, p. 47) suggested the importance of "dropping questions altogether for a time" when, in the interviewer's opinion, the child consistently avoids addressing the point of the question. Rich also noted the following:

> For young children or for dull children it is important to put questions in concrete terms; for example, not "what are your hobbies?" but "what do you do in the evenings when you're not at school?" They may believe that "hobbies" means "handicrafts" and reply "I don't have any," although in fact they do all sorts of things.

Techniques for Dealing with Communication Blocks

Persons often block communicating because they wish to avoid a certain topic. In such cases, the interviewer must be sensitive to painful areas and should approach cautiously from several directions. Indirect questions are useful in these situations. Another cause of communication block is one's inability to express oneself. Sometimes a neutral question, such as "explain what you mean," "I don't quite understand," or "tell me more about it," helps to clarify meaning. It is sometimes helpful for the interviewer to apologize for not understanding. This takes some of the burden for communicating away from the child. A third cause of communication block may be the environment of the interview room. It is helpful to consider the following questions:

> What do I want the child to hear?
> Does this room enhance his hearing this, or does it encourage his hearing something else?
> What do I want the child to do as he enters?

So often, in an instructional setting, thought is not given to establishing rapport before getting to the purpose of the interview. The importance of establishing rapport is built into the training of school psychologists, sometimes for physicians, and not very often for classroom teachers. Also, we become so used to our own interview settings that we do not always think about how they look to—and affect—an interviewee. For example, the child who is anxious about mathematics to begin with, when confronted with an array of mathematics tests or arithmetic manipulatives upon entering the interview setting, may freeze. This is analogous to a doctor's office in which strange and frightening instruments are in full view, yet the doctor assures the frightened child that he will not hurt him during an examination. The doctor may then be puzzled why his reassurances do not allay the child's evident fear. When engaging in a diagnostic mathematics interview, therefore, keep materials that might overwhelm the child out of sight, and bring into view a few at a time, as needed.

A familiar setting is helpful for establishing rapport. A desk or table similar to that in the child's classroom is preferable to a strange living room setting. The living room setting might establish incongruencies between "teacher figure" and "living room" environment; this might worry the child rather than put him at ease. A child often feels safer if there is a table or desk at least partially between himself and the interviewer, especially if the interviewer is a stranger to the child.

In some cases, the interviewer may want to give the child an opportunity to look around. You may greet the child by name and then busy yourself for a few moments to allow the child time to look you and the room over. Of course, if you are dealing with an aggressive, behavior-disordered child, this approach might be inappropriate. (For such a child, rapid initiation of a controlled situation might be necessary to prevent his tearing up the room.)

Some interviews will occur in the child's classroom setting. In this situation, two points should be considered:

1.   Place the child in a part of the room where others will not disturb the interview. Sometimes the child's own desk best accomplishes this because all remains familiar to other children in the room.
2.   Avoid failure-related settings. Move to another desk or table at a pleasant part of the room, or stay at the child's desk if he does well when situated there.

Other communication blocks include the following:

• Differences may exist in vocabulary and usage of words due to cultural differences between the interviewer and child. A word may signal hostility to one person and may be used as a positive term to another, such as "he's a mess" or "that's tough."
• A child's antagonism toward a parent or toward another teacher may be transferred to the interviewer.
• The child may be "testing limits" to see how far he or she can push the interviewer.
• A child may be responding to real negativeness on the part of the interviewer.

In summary, techniques for dealing with communication blocks may need to be modified or adapted in relationship to one or more of the following characteristics of the student: age, intellectual development, cultural and socio-economic characteristics, number of interviews child has experienced, history of interviewer–child relationships, and specific handicap, e.g. language or behavior disorder.

Further Suggestions and Comments

Spontaneous remarks often provide important information. When working with children, we must recognize that their spontaneous actions and behavior often substitute for spontaneous remarks (Rich, 1974).

Behavior after the interview should be observed, because a child who is anxious and noncommunicative during the interview often loosens up after he thinks the interview is over. The teacher may build this situation into the interview by putting materials away, walking over to some object in the room and remarking about how attractive it is, or going to the window and making some remark about the weather—anything to signal the end of the interview. The child may then start to talk about things he previously avoided that might provide insight for the interviewer.

Look for contradictions between what the child says and what he does. Sometimes giving the child materials to work his hands with, such as clay, provides opportunity to observe verbal–nonverbal discrepancies. For example, the child may be describing how much he likes his arithmetic teacher while simultaneously pounding a clay figure of a person flat on the table.

Be aware that new points raised by the child after a comparatively long period of silence may be more closely related to a previous statement than one might expect. The child may have been incubating ideas and forming relationships in his mind, and you may be hearing only the endpoints of his thinking.

Be on the lookout for recurrent themes, as these may show preoccupation with one problem situation or person. A recurrent theme sometimes is not apparent until you go over your notes or listen to a recording of an entire interview (or series of interviews).

Interviewers have different opinions about note taking during an interview. Some say note taking inhibits the child and interferes with the interviewer's attention. We believe the value of the information obtained usually outweighs the negative aspects of either note taking or recording the session (videotapes or audiotapes). Often, things are spotted when reviewing notes that were completely missed during the session; information may be lost to fading memory.

Be aware of a child's strengths as well as weaknesses. Do not focus only on what she *cannot* do. It is important to ask yourself, "Why can she perform *some* problem-solving tasks and not others? Why did she learn to rename in multiplication and not in addition? Why is her comprehension good when asked to explain main points of a story, but not when asked to tell what is asked for in an arithmetic word problem?"

Do not confuse observation with inferences. For example, it should not be inferred that a student understands place value if he does not raise his hand when the teacher says, "Any questions on place value?" A child who does not understand a topic may not indicate lack of comprehension for many reasons—shyness, fear, or *not realizing* he does not understand. Instead, in this example the teacher should *observe* the student's performance on tasks dealing with place value.

Note the child's appearance and the way in which he moves. What about his relationships with other children and with adults? What are characteristics of his verbal behavior? What about his general mood and his interaction with his environment? How about eating and sleeping habits? What about his behavior in school-related activities, including academic subjects, as well as in other activities (art, music, physical education, or playground games)? See Appendix B for an interview checklist that summarizes points to note as part of the interviewing process.

Summary Thought Questions and Activities

I. Play a tape-recorded interview that you have conducted to a group of colleagues who are also working in a similar capacity. Allow approximately 2 hours for every half hour on the tape, so that you can stop the tape whenever one of the group wishes to comment or question the interviewer. The interviewer should be able to justify his interviewing behavior—remarks made, silences allowed to develop, vocabulary used, and subjects discussed.

II. Stop a taped interview and ask others to predict what will happen next, or how they would proceed.

III. Evaluate whether your interview worked with regard to the following criteria:
    A. The information collected is accurate.
    B. Behavior of the child changes in the desired direction.

IV. Discuss alternate criteria that might have been selected in item III above, and explore effects of implementing each of these.

Suggested Readings

Adams, J. S. *Interviewing procedures.* Chapel Hill: University of North Carolina Press, 1958.

Fenlason, A. F. *Essentials in interviewing.* New York: Harper, 1952.

Kahn, R. L., & Cannell, C. F. *The dynamics of interviewing.* New York: Wiley, 1961.

Luria, A. R. [*Cognitive development—Its cultural and social foundations*] (M. Cole, ed., and M. Lopez & L. Solotaroff, trans.). Cambridge, Mass.: Harvard University Press, 1976.

Reference

Rich, J. *Interviewing children and adolescents.* London: The MacMillan Press, Ltd., 1974.

Endnote

1. The following illustrates an interview situation in which the real problem was obscured.

A 5-year-old boy did not use speech to communicate. His mother brought him to a hospital for help. She also brought along her younger child—a noisy, crying 10-month-old boy, whom she constantly wheeled back and forth in his stroller as she attempted to answer questions.

The first of the interviews, conducted by a pediatrician whose primary attention was directed to four medical students and a technician who was videotaping the session, ended in the youngster becoming frustrated to the point of crying in terror as he sat on his mother's lap. The pediatrician's fact-finding brought out, among other things, that the birth was difficult, the maternal grandmother thought the boy was retarded (he obviously was not), and the father had been shot in the head 3 months earlier and was confined to an upstairs room because of total paralysis.

The pediatrician attempted to examine the child, and her efforts to approach him triggered near-hysteria in the child. The mother was told by the pediatrician not to comfort the child as this would reinforce his hysterical behavior. For the next 15 minutes, one mother, one pediatrician, two educators (including one of the present authors), four medical students, a videotape technician, a bewildered 10-month-old infant, and a screaming 5-year-old sat around a table. The child finally succeeded in terminating this interview.

The two educators then accompanied this family to the second interview, conducted by a psychiatrist. The same questions were repeated:

> Why are you here?
> Did he ever talk?
> Does he talk to you?
> Was there anything unusual about the birth?
> Do you work?
> When did he begin to walk?
> When did he first sit up in the crib?
> Does he always try to get attention by falling off chairs? (Child continuously came near falling.)

In this interview, the condition of the father never came out. The mother began to wonder when the fact-finding would end. The little boy responded to the mother's directions, including a helping routine with the baby ("get the baby's bottle"); "here's a piece of gum for you"; and "go for a walk with the lady" (responding to an offer by the author's educator friend), which he did. The interview ended with an appointment being made for their return the next week. They never returned.

Instead of disclosing facts that may have benefited the child, this interview produced superficial and stereotypical results. The hospital personnel were so intent on obtaining multiple-choice-type background information that they never listened to what the mother wanted to say. She had come to obtain professional help for her son, and all she incurred was frustration.

Mainstreaming and the Individualized Mathematics Program (IEPM)

14 Implementing Public Law 94-142, concerning children and young adults with special educational needs (the handicapped and the gifted), involves mainstreaming and development of Individualized Education Programs. In this chapter we will focus particularly on developing individualized mathematics educational programs (IEPMs).

Mainstreaming

Reynolds and Birch (1977) listed eight events or developments that provide historical background for mainstreaming. These changes, which affect the roles of teachers and principals, include legal decisions, public school obligations, and parental rights.

1. The wave of judicial pronouncements concerning the right to education, right to appropriate education in the least restrictive environment, and right to participation in placement decisions.
2. The virtual "shutdown" on school exclusions, expulsions, and suspensions.
3. The return of many persons from residential institutions to the community, resulting in the obligation of local schools to serve many seriously handicapped students.
4. The launching by local schools of programs for persons formerly believed to be uneducable.
5. The rapid erosion of the traditional boundaries between regular and special education.
6. The emergence of new support or indirect roles for special education teachers that involve teaming with regular educators.
7. The requirements that parents be notified before any special "diagnosis" or assessment of their children is undertaken in educational planning.
8. The serious objections by many parents and minority groups to the prevailing systems for testing, classification, and placement of children in special classes in the schools. (pp. 3–5)

These eight developments point to a broader inclusion of children with special needs in the *mainstream* of school and community life. Mainstreaming thus implies a shift from isolating students with special educational needs on account of their problems to a philosophy of building on their strengths.

212

Public Law 94-142 was enacted by Congress in November of 1975. The major purpose of the act is stated (Federal Register, 1977) as follows:

> It is the purpose of this Act to assure that all handicapped children have available to them . . . a free appropriate education which emphasizes special education and related services designed to meet their unique needs to assure that the rights of handicapped children and their parents or guardians are protected, to assist states and localities to provide for the education of handicapped children, and to assess and assure the effectiveness of efforts to educate handicapped children.

P. L. 94-142 involves six major rules or regulations. These include the following:

1. The principle of *zero reject*, which requires that all handicapped children be provided a free and appropriate public education. States had to provide this educational opportunity to all handicapped children, ages 3 to 18, by September 1, 1978, and for ages 3 to 21 by September 1, 1980.
2. *Nondiscriminatory evaluation* procedures must meet the following critera:
   a. Tests and other evaluation materials:
      1. are provided and administered in the child's native language or other mode of communication, unless it is clearly not feasible to do so;
      2. have been validated for the specific purpose for which they are used; and
      3. are administered by trained personnel in conformance with the instructions provided by their producer.
   b. Tests and other evaluation materials are tailored to assess specific areas of educational need and are not merely designed to provide a single, general intelligence quotient.
   c. Tests are selected and administered so as to ensure that the test results accurately reflect the child's aptitude or achievement level or whatever other factors the test purports to measure, rather than reflecting the child's impaired sensory, manual, or speaking skills (except where those skills are the factors which the test purports to measure).
   d. No single procedure is used as the sole criterion for determining an appropriate educational program for a child.
   e. The evaluation is made by a multidisciplinary team or group of persons, including at least one teacher or other specialist with knowledge in the area of the suspected disability.
   f. The child is assessed in all areas related to the suspected disability, including, where appropriate, health, vision, hearing, social and emotional status, general intelligence, academic performance, communicative status, and motor abilities. (Federal Register, 1977, pp. 42496-97)
3. Placement of children in the *least restrictive environment* must be based on the student's IEP goals and objectives. These goals and objectives are devised after a careful analysis of all evaluation data. Following this review, a determination is made regarding which academic areas need modifica-

tion. If a student's special learning needs are incompatible with the curriculum of the regular classroom for a specified subject, then mainstreaming would not be the *least restrictive environment*. In fact, in this instance, the regular classroom would be an inappropriate placement.

4. *Due process* procedures—regarding identification, evaluation, and placement, or for challenging decisions (such as when parents and educators cannot agree on IEP goals and objectives)—are stated in P.L. 94–142.
5. *Parents must be included* in the development and approval of educational policy at the local and state levels. Parents also have access to educational records and information on their child.
6. *Development of an IEP* for each handicapped student must include the following:
    a. documentation of the student's current level of educational performance;
    b. annual goals expected to be attained by the end of the school year;
    c. short-term objectives, stated in observable terms;
    d. designation of special education and related services that will be provided to the child;
    e. the extent ot time the child will participate in the regular education program;
    f. projected dates for initiating services and anticipated duration of these services; and
    g. evaluation procedures and schedules for determining mastery of short-term objectives at least on an annual basis.

Developing an IEPM

The actual IEPM, which comprises the mathematics goals and activities of the IEP, is written by members of the IEP committee—most likely the regular classroom teacher and the special education teacher, who are most involved (or will be) in implementing the goals and objectives of the IEPM. Specific activities of the IEPM committee include the following tasks:

1. Specify child's level of performance in mathematics.
2. Decide upon and prioritize long-term (annual) goals in mathematics.
3. Write a task analysis for each long-term goal (these comprise the short-term goals of the IEPM.
4. Translate into behavioral objectives each step on the task analysis for a particular long-term goal.
5. Specify evaluation procedures for each goal.
6. Determine placement, related services (counseling, medical, recreation, and so forth), and extent of time in regular classroom.
7. Obtain agreement regarding IEP from all members of the IEP committee (teachers, administrator, school psychologist, counselor, representative from public health agency, medical personnel if necessary).

Functions of the IEPM Committee

The committee should obtain the following background information on the referred child:

1. cognitive functioning (for example, intelligence, adaptive behavior, and present achievement); physical functioning (for example, vision, hearing, fine and gross motor coordination, and general health); social-emotional functioning (for example, relationships with peers and adults, group participation, adaptive social behavior, and usual classroom behavior); and parent–teacher interaction if relevant
2. summary of medical reports, including relevant medical history
3. information on the family and home environment, including the child's behavior in the home
4. educational history, including what the child was able to learn, what was always a trouble area, and the rate of development in various skill areas. This information may be obtained from, previous teachers—regular classroom, physical education, music, art, or special education; from the school counselor; from an administrator; or from cumulative records, if child was recently moved from an institution for the developmentally disabled to a foster home and will now be placed in a public school setting.

The IEP committee must meet within 30 days from the date of determination that a student with handicaps is in need of special services. The IEP committee should recommend appropriate placement for implementation of the IEP. Members also are responsible for devising strategies for evaluating whether IEP goals are being accomplished. These goals, as well as the short-term objectives, must be revised at least annually, according to P.L. 94-142.

Annual Goals

The following suggestions for specifying annual goals were presented by Turnbull, Strickland, and Brantley (1978):

1. Goals chosen should be practical and relevant to the social and vocational needs of the student.
2. Annual goals should reflect the most essential needs of the student.
3. Goals should be sequenced from simple to complex and prerequisite skills taught prior to ones that are more complex.
4. The IEP committee must make an estimate of how much instruction can occur within a given amount of time and must determine annual goals accordingly.
5. Many goals will take more than one year to accomplish. The IEP should therefore be considered an ongoing plan that will enable students with special educational needs to achieve their maximum potential within the time given upon leaving school. (pp. 152–154)

Example of Annual Mathematics Goals

Annual or long-term goals are global and comprise a hierarchy of short-term, prerequisite goals. Examples of long-term goals will be described for Carlton, a student whose case study is presented below. (All data are fictitious.)

Case Study: **Carlton**

Referral Form:
NAME Carlton Jones   AGE 8 years 9 months
SCHOOL Main Street Elementary   GRADE 3   SEX M
PARENTS/GUARDIAN Mr. & Mrs. Harry Jones
ADDRESS Sunny Way Trailer Park   PHONE none
REFERRED BY Ms. White   POSITION 3 grade teacher
REASON FOR REFERRAL Evaluation by multidisciplinary team medical consultation

---

PROBLEM AREAS

Acculturation and language: no problem
Health: well developed, attractive, no problem
Hearing: appears normal
Vision: appears normal
Speech: difficulty in oral reading, poor intonation
Perceptual–motor performance: poor drawing, messy writing, good at sports
Academic achievement: 3rd stanine in mathematics, 5th stanine in reading comprehension, problems in reasoning, uses trial-and-error problem-solving strategies.
Social–emotional behavior: short attention span, distractible, hyperactive, withdrawn, relates poorly with adults and peers.
Parental involvement: mother takes him to the doctor when he "gets nervous," says she "can't control him"; father came to a morning parent–teacher conference smelling of alcohol
Other comments: teacher would like neurological evaluation and wonders about possibility of medication that might control Carlton's hyperactivity

SUMMARY OF NEUROLOGICAL REPORT

Carlton was brought to the doctor's office by his mother. An older sister and a younger brother, Lenny, accompanied him. The younger brother, age 3, also was to be examined that day with similar behavior problems. The two older sisters were said to be fine and doing well in school. The mother, 38, had a 10th grade education. The father, 46, obtained a high school equivalency diploma. He was a chronic alcoholic and recently had lost his driver's license because of drunken driving. The mother said that several of her own sisters have been in foster homes and at detention school because they would skip school and they were discipline problems. The mother stated that one of her sisters was "running around with another man after being married and [would come] to her house often swearing and not being a good example for Carlton."

The mother described Carlton as "having a mind of his own, pesters his brother and sisters, fights, does not like school, is very picky about food and does not want to eat what she prepares." She stated that she could not discipline him, and referred to him as a "devil at home and an angel in the street." She said there were no complaints from the school where he just finished third grade and that he behaves fairly well. He got

C's and B's in school and supposedly was best in mathematics. She said he was basically healthy. His mother was expecting to find some sort of a "nerve pill" that would calm him down and stop him from "being constantly on the go." He went to bed at 9 o'clock and got up at 8:30 in the summer when there is no school.

The pregnancy and delivery were normal. The birth weight was 8 lb., 12 oz. Carlton's development in walking and talking was average. At 3 months of age, he had had some sort of blockage according to the mother, but required no surgery.

Carlton was well developed, attractive, with no midline (spinal) defects, birth marks, or other unusual abnormalities in external configurations. He was very shy and sat in his chair quietly, just smiling and hardly said anything—he nodded his head to indicate yes or no. He admitted to doing most of the things his mother described and shrugged his shoulders when asked why he would do such things.

His draw-a-man figure was done carelessly. He first did a stick man; when asked to elaborate, he drew a boy figure, to which he added one detail after another as encouraged to do so. The total score was just above a 7 years of age norm. On the similarities subtest of the WISC-R (Wechsler Intelligence Test for Children—Revised), he made conceptual associations but needed concrete clues and encouragement. He read third grade material well, but sentence cadence was not very good and he did not make the proper intonation at the end of the sentences. He could do two-digit additions and subtractions without problems. On the Bender Gestalt he copied figures no. 2, 4, 6, 7 and 8 without any problem and quite correctly. His head circumference was 56 cm, which is normal.

His general physical examination was unremarkable. He walked well on his heels and toes, in a straight line; and hopped and skipped without problems. He could not, however, skip on a small spot repeatedly. He had no tremors or other involuntary movements. He was fair with the sticks-between-the-fingers test, chiefly because he was in a hurry. He was able to drop only one stick at a time. His reflexes were bilaterally equal and normal. Finger-to-finger and finger-to-nose maneuvers were done normally, and the sensations were normal. His cranial nerves show normal development.

In conclusion, Carlton was normal neurologically and intellectually (within average range). It is interesting that he functioned fairly well in school, though according to his mother "he does not talk much there and he sits in a corner without participating much." The behavior problem seemed to be a reaction to his home environment, in which the male figure was far from optimal and the mother's skills to handle him appeared limited. Since the basic problem appeared to be environmental and educational, Carlton did not warrant stimulant therapy at this time; drugs would not solve the problems. He was given an appointment to return with a school report in his new grade. At that time a decision would be made regarding medication. If his behavior were to warrant it, a short period of medication might be considered. At that time, an assessment by his teachers of his performance and behavior changes would give more data for evaluation than just the mother's observation during the summer.

TEACHER'S SUMMARY OF INTERVIEW CHECKLIST (APPENDIX B)

Appearance: appropriate dress; facial expression often placid, vacant, but happy; physical health is good, weight appropriate

Movement: observed—quick, well-coordinated, restless; inferred—not always purposeful, mostly meaningful, moves freely

Emotional: expression is usually pleasant, anxious, suspicious; mood fluctuates and sometimes overreactive

Orientation: spatial—knows where he is, knows how to get from one place to another; temporal—knows time of day, on time usually

Verbal: talks a little to teacher, a lot with younger children, answers only when talking with teachers, sudden silences, stays on topic for appropriate time, appropriate level of vocabulary, limited syntax

Interpersonal relationships: an isolate much of time, picks friends much younger, sometimes participates in group, sometimes friendly to familiar adults

Interactions in general: rebellious of rules, instructional program—just goes along, needs direction, handles frustration poorly, distractible during lesson, preoccupied, offers few ideas, flexible, mundane, needs structure, needs quick closure

TEACHER'S SUMMARY OF IS-SLDM CHECKLIST (APPENDIX C)

Reasoning: Strengths—can sequence, displays logical errors in computations, uses cues correctly when involving concrete activities (for example, puzzles, mazes, memorizing basic facts in arithmetic); *weaknesses*—inability to make choices and decisions, does not connect cause–effect relationships, does not make appropriate inferences and draws inappropriate conclusions, misuse of syntax.

Problem solving: Strengths—completes most tasks on time; *weaknesses*—uses trial-and-error approach, does not organize tasks

Orientation: Strengths—indicates knowledge of spatial relationships, shows awareness of temporal relationships; *weakness*—reverses place value notation (e.g., 32 = 2 tens, 3 units) computes + , ×, - algorithms in left-right directions

Motor performance: Strengths—not clumsy, good gross motor coordination; *weaknesses*—excess tongue movement under stress, excess body movement, messy writing, immature eye–hand coordination, displays expressive language disabilities

Attention: distractible

Perception: seems OK

Affect: ambivalent regarding self-concept; displays inappropriate level of anxiety for task difficulty

TEACHER'S SUMMARY: DEVELOPMENTAL MATHEMATICS CURRICULUM SCREENING CHECKLIST (APPENDIX D)

I. *Basic relationships*: Strengths—prenumber relationships, topological relationships, arbitrary associations (words); *weaknesses*—symbols.

II. *Lower-level generalizations*: Strengths—duality, set relationships, seriation, binary judgments; *weaknesses*—multiple judgments

III. *Concepts* : Strengths—number, geometry, time, money *weaknesses*—size and amount

IV. *Higher-level relationships*: Strengths—transformations, set operations and relationships, binary operations, *weaknesses*—inequalities, equivalence relations, mathematics language

V. *Higher-level-generalizations*: Strengths— basic +, ×, ÷facts; applied mathematics (time/calendar, clothing size, temperature, economics) *weaknesses*—basic ÷ facts, axioms, place-value notation

Carlton's goals in mathematics are as follows:

I. Carlton will write dictated numerals in expanded notation to show place value of each digit.

II. Carlton will compute addition algorithms with renaming.

III. Carlton will use well-planned problem-solving strategies in tasks.

IV. Carlton will correctly apply cues from mathematical language, such as inflectional endings and plurals.

Short-Term Goals and Behavioral Objectives

Short-term goals are the components of the annual goals—the subtasks, the necessary prerequisites that one must know before understanding an end goal. A short-term goal may be translated into *behavioral objectives* that tap different types and levels of competence—for example, allowing the student to show what he knows by pointing, drawing, writing, saying, or representing with concrete materials, pictures, or symbols. Both short-term goals and their components, behavioral objectives, must be placed in the most optimal sequence for learning and prioritized for the student.

It is sometimes helpful to analyze a long-term goal into its component objectives rather than attempting to jump directly to performance objectives. For example, it is helpful for the teacher to ask, "What does the student have to *know* or *understand* before he can use place value?" Then, list all prerequisites as you analyze the task, *using place value*. It is a good idea to write one prerequisite per index card. Then you can more easily arrange them into a learnable sequence.

When one performs a task analysis, salient prerequisites normally considered as task components should be included. For example, it would be a basic omission to develop a task analysis for *using place value to write numerals* and leave out *writing the digits 0–9*. Following is one approach to analyzing this particular task (Annual Goal I) for Carlton:

Long-Term Goal I: to write dictated numerals in expanded notation to show place value

1. can count 0–9
2. can name face value of a digit
3. can identify the larger (or smaller) of two numbers
4. can write the digits 0–9
5. understands many-to-one relationship
6. can multiply factors to obtain products
7. knows the names of multiples of 10 through 900
8. knows the verbal names for numbers
9. knows that place value is a product
10. knows the names of the places to the left of the units place
11. knows that place values increase by powers of 10 to the left
12. can name place value of a digit within a multidigit numeral
13. knows that the value of a digit within a multidigit numeral has the value of its face value times its place value, so its value is a product
14. can name concrete representation of a two-digit numeral
15. can write a two-digit numeral from a concrete representation
16. can write a two-digit numeral in standard form
17. can write a two-digit numeral in expanded form
18. can write a dictated two-digit numeral in expanded form

Following are Carlton's Long-Term Goal I; three of the prerequisite short-term goals; and, for each of these, four behavioral objectives. The teacher may wish to use the *Developmental Mathematics Curriculum Screening Checklist* (see Appendix D) as a guide for identifying subtasks.[1] Once a long-term goal has been analyzed into its subtasks, the same structure may be reused for several students.

Long-Term Goal I

Carlton will write dictated numerals in expanded notation to show place value.

Short-Term Goals (STG)	Behavioral Objectives
1. can write the digits 0–9	a. to say the digits 0–9 b. to point to the digits 0–9 c. to trace the digits 0–9 d. to copy the digits 0–9
2. knows that place value is a product	a. to show many-to-one relationship b. to compute powers of 10 c. to name the powers of 10 through hundreds d. to name the place values in a base 10 numeration system
3. can write a two-digit numeral from concrete representation	a. to show 0–9 on the R-K counting boards b. to write the numerals 0–9 that are shown on single R-K counting board that represents the units place c. to show from 0–19 on the R-K counting boards d. to write the numerals 10–99 that are shown on the R-K counting boards (see section 9.3.0, especially IS 9.3.2)

The next step is to decide on appropriate behaviors the student may use to show that he understands a particular prerequisite. At this point it is helpful to develop a "table of specification" to serve as a structure of the content and the behaviors you will use to state the observable objectives (see Table 14.1).

Another teacher might very well have come up with different short-term goals and behavioral objectives. This is to be expected and is not a problem, so long as there is consistency in sequencing simpler skills before skills of which they are components. For those who find it difficult to write behavioral objectives, keep in mind that the verb used should represent something you can *see the student do*. Examples of such verbs are as follows: draw, point to,

Table 14.1. *Table of Specification for Carlton's Long-Term Goal I*

Content	Say	Point to	Trace	Copy	Compute	Show	Name	Write
			Behavior					
0–9	STG-1a	STG-1b	STG-1c	STG-1d		STG-3a		STG-3b
0–19						STG-3c		
10–99								STG-3d
Many–1 corr.						STG-2a		
Powers of 10					STG-2b		STG-2c	
Place values							STG-2d	

record, say, state, trace, write, compute, describe, discuss, translate, explain, construct, measure, select, show, exchange.

Evaluation of Goals and Objectives

P. L. 94–142 requires three aspects for evaluating IEP goals and objectives. These are as follows (Turnbull, Strickland, & Brantley, 1978, p. 170):

1. Determine objective criteria, e.g., 80% correct or 90% accuracy.
2. Determine appropriate evaluation procedures, e.g., complexity of task (recall or analyze), mode of representing content (concrete, picture, with symbols), time allotted for instruction, and criterion requirement.
3. Determine evaluation schedules, e.g., daily evaluation as well as monthly and annually.

IEP Formats

There are several forms that are typical of IEP formats. Two of these that seem appropriate for IEPMs are mentioned here. The *IEPM Implementation Plan* form, illustrated below for Carlton, is particularly appropriate for the teacher who will carry out the instructional strategies with the student. The *IEPM Total Service Plan* format presents a broader plan which may be sufficient for some students; this *Total Service Plan* is used in the IMP's presented in Appendix F. It should be noted that the column labeled *IS* stands for *Instructional Strategy*. This column replaces the usual IEPM column labeled *Materials*.

IEPM for Carlton

Student: Carlton Jones   *Subject Area*: Mathematics
Level of Performance: poor interpersonal relationships; can perform tasks involving (a) basic relationships; (b) lower-level generalizations *except* multiple judgments; (c) concepts *except* size and amount; (d) higher-level relationships *except* inequalities, equivalence relations, and mathematics language; (e) higher-level generalizations *except* basic ÷ facts, axioms, and place-value notation.

Annual Goals:
I. Will write dictated numerals in expanded notation to show place value of each digit.
II. Will compute addition algorithms with renaming.
III. Will use well-planned problem-solving strategies in mathematics tasks.
IV. Will correctly apply cues from mathematical language, e.g., inflectional endings, plurals.

Short-Term Goals	Behavioral Objectives	IS	% Accuracy of Task	Date Achieved	Persons Responsible
Annual Goal I					
Write digits, 0–9	To say digits 0–9	5.1.1, 5.1.4	90%		Ms. White, 3rd grade teacher
	To point to digits 0–9	5.1.2	"		"
	To trace digits 0–9	5.1.3	"		"
	To copy digits 0–9	5.1.3	"		"
Knows that place value is a product	To show many–one relationship	6.1.3	"		"
	To compute powers of 10	5.1.9	"		"
	To name powers of ten-hundreds	5.2.4	"		"
	To name place values in base ten	6.1.2	"		"
Can write two-digit numeral concretely represented	To show 0–9 on R-K counting boards	9.1.5	"		Resource Teacher for Behavior-Disordered
	To write 0–9 from R-K board	9.1.5	"		"
	To show 0–19 on R-K boards	9.3.1	"		"
	To write 10–99 from R-K boards	9.3.3	"		"
Annual Goal II					
Knows place value	See behavioral objectives for Annual Goal I				
Knows basic addition facts	To compute basic addition facts	9.1.5	90%		
Applies appropriate axioms	To show commutative property for +	9.2.2	"		Ms. White
	To show associative property for +	9.2.3	"		"
Knows how to arrange digits in vertical + algorithm	To copy vertical addition algorithms	10.1.1	"		"
	To write + algorithms in vertical form	10.1.1	"		"

222

Short-Term Goals	Behavioral Objectives	IS	% Accuracy of Task	Date Achieved	Persons Responsible
Knows how to process + algorithm	To compute using vertical + algorithm	10.1.1	80%		BD Teacher
	To use historical algorithm	10.1.1	"		"
Can add with renaming	To make 10–1 exchange	9.3.3	"		Ms. White
Operates according to + sign	To do what operation sign indicates	8.3.4	90%		"
Knows that only one digit is written per column	See Goal I				
Know that only similar place values are added	To classify like things	7.1.1	"		"
	To add like place values	9.4.0	"		"
Can write + algorithm legibly	To write digits legibly and in their proper place in algorithm	5.1.3	"		"
Annual Goal III					
Can make appropriate inferences	To infer correctly	5.1.9, 5.1.10, 5.3.2	80%		Ms. White & BD Resource Teacher
Can organize task to facilitate its completion	To organize tasks	10.1.4	"		"
Use planned approach to solving tasks	To use well-planned problem solving	5.1.6–8 5.1.11–12 5.2.4 5.3.1–2 5.4.1	80%		"
Annual Goal IV					
Can use mathematical language	To use correct syntax	8.4.4, 8.4.5	80%		Ms. White & BD Resource Teacher
	Not to reverse word meanings	5.2.1, 5.3.1–2 5.4.1	"		"
	To use cues from inflectional endings correctly	8.4.6			

Summary: Carlton

The IEPM committee consisted of Ms. White, Carlton's third grade teacher; the school's special education teacher for children with behavior disorders (BD); the school principal; the school psychologist, who had spoken with both the pediatric neurologist and the family's physician and was presenting their findings; and Carlton's mother and father.

The goals in the area of social interaction took priority over mathematics goals. However, it was believed that a focus on Carlton's strengths in mathematics that became apparent from the checklists would be a good starting place to reach him. By providing him with some success experiences in mathematics, it was hoped that he would be reinforced and would become interested in tasks he could do well.

It was agreed that Carlton would work with the BD teacher for one hour each day in a small group, and this interaction would include some short-term goals in mathematics. He seemed to calm down and to work more carefully when told what to do step by step; this could be done by the BD teacher.

Summary Thought Questions and Activities

I. Write three long-term goals for a handicapped or talented student.
   A. Select one of these long-term goals and write three short-term objectives that are components of the long-term goal.
   B. Select one of these short-term objectives and write three behavioral objectives that reflect the short-term goal.
II. Outline the crucial points from Carlton's case study that show his patterns of behavior, his learning strengths and weaknesses and inconsistencies in data accumulated on Carlton. Compare your analysis with the objectives stated in his IEPM.
III. Select a child in your class or neighborhood and evaluate her or him by using one of the checklists in the appendices. Describe your predictions of this child's mathematics performance. If possible, verify your impressions by obtaining feedback from the teacher or parent.

References

Federal Register. Washington, D.C.: U.S. Government Printing office, August 23, 1977.

Reynolds, M., & Birch, J. Teaching exceptional children in all America's schools. Reston, Va.: The Council for Exceptional children, 1977.

Turnbull, A. P., Strickland, B. B., & Brantley, J. C. Developing and implementing individualized education programs. Columbus, Ohio: Charles E. Merrill Publishing Company, 1978.

Endnote

1. There are commercial materials available that may serve as a guide for analyzing mathematics tasks: Noonan-Spradley Diagnostic Program of Computational Skills, P.O. Box 78, Galien, Michigan 49113; Allied Education Council Distribution Center, 1970; Lola J. May & Vernon Hood, BASE (Basic arithmetic skills

evaluation) (1973), Media Research Associates, 1735 23rd Street S.E., Salem, Oregon 97302; *Fountain Valley Teacher Support System in Mathematics* (1972), Richard L. Zweig Associates, Inc., 20800 Beach Boulevard, Huntington Beach, California 92648; see also scope and sequence charts that form the structure for basal mathematics series and teachers' methods texts in mathematics.

"How Can I Help My Child?"
Changing Roles of Parents and Teachers

15 How many of you remember your parents asking one of your mathematics teachers, "How can I help her? Isn't there anything I can do to help?" Have you ever been in this circumstance with your own child? Has a parent ever asked this of *you*? What was the reply?

New roles for regular classroom teachers, special education teachers, and parents are emerging as a result of P.L. 94–142. One of the major changes is the cooperative development of educational goals for a child by the IEP committee, which includes the parent or guardian.

Parents now have legal access to educational records and information concerning their child and must be involved in shaping the educational objectives for their child. Implementation of these objectives may depend upon bridging activities in the home or community. For example, if a classroom objective is "to tell time to the precision of a minute," the parents or guardians must provide time-telling activities so that this becomes a useful tool rather than a splinter skill. Using school-learned tasks in the home helps the child to generalize to real-life situations and provides opportunities for practice. For a cognitively slower child, mathematics must serve a useful purpose—a life need. Given the number of years he or she will be taught, and realizing that a slow learning rate is involved, those who select the mathematics such a child is exposed to must provide *relevant* mathematics. This is also true for a child with low energy levels, short attention span, or any other special educational need. Selection of relevant and learnable mathematics will be guided more and more by the child's special and unique needs, the teacher's professional competence, and the parents' values and judgments.

The renewed emphasis on differential mathematics instruction encourages diagnostic teaching. Analyzing mathematics goals to identify prerequisites missing from a child's repertoire—a major component of diagnostic teaching—is essential in developing an individualized mathematics program (IEPM).

Developing IEPMs will have some very helpful benefits. With regard to parent–teacher cooperation, these benefits include the following:

Teachers will suggest activities (for example, selections from the ISs presented in Chapters 5 through 12) that parents can use to help their child with at home. Activities may also result from the teacher's task analyses of mathematics content, necessary for writing IEPM short-term objectives. Thus, when a parent asks, "What can I do to help my child in mathematics?" the teacher will have specific guides for enlisting parents' help in practice and transfer activities.

- Parents will gain insight into generic factors that influence how their child learns mathematics. They will come to appreciate their child's efforts as they observe, in a first-hand interaction, their child's special education needs.
- Parents will become aware of their child's strengths as instructional strategies facilitate strengths that may have been neglected previously under traditional instruction.
- Parents will be able to provide teachers with another view of children's strengths and special needs. For example, a cognitively slower child may look like a loser in the classroom milieu but may be a winner in the home, community, and life outside of the classroom. A child frustrated by mathematics tasks that are too complex may be distractible and disruptive in school, and yet display no such characteristics in settings that allow him some success.
- Emphasis on interdisciplinary educational strategies will become more prevalent. Just as boundaries and barriers between regular and special education are being eliminated, barriers between the school, the home, and the community are giving way to a team development of educational policy and goals for a child.

Parents' Rights Under P.L. 94-142

P.L. 94-142 requires that before a child is referred for consideration for special services, the parent or guardian must be informed. Parental consent must be obtained prior to formal evaluation, and the parents must be informed of their rights.

The law requires that parents be informed, by written notice within a reasonable time, that the school intends to evaluate their child. Parents must be informed of results of their child's evaluation and of how they will be involved in developing their child's IEP. The law states that only children who will receive special services must have an IEP developed for them. However, *all* children referred must have their evaluation results communicated to the parent. The notice of results of the evaluation that is transmitted to the parents must include the following (*Federal Register*, August 23, 1977, p. 42495):

A full explanation of all of the procedural safeguards available to parents;
A description of the action proposed or refused by the agency, and explanation of why the agency proposes or refused to take the action, and a description of any options the agency considered and the reasons why those options were rejected;
A description of each evaluation procedure, test, record, or report the agency uses as a basis for the proposal or refusal;
A description of any other factors which are relevant to the agency's proposal or refusal.

227

Summary Thought Questions and Suggested Activities

1. Describe, in one or two paragraphs, the relationships among a child, his or her parents or guardian and the school in developing an IEPM.
2. Role-play with a colleague:
   The colleague has a 9-year-old child in your class who is developing cognitively at a rapid rate. This child cannot subtract with renaming. The colleague asks, "How can I help my child learn to subtract with borrowing?"
   A. Show the parent (colleague) a task analysis of subtraction with renaming that you have developed.
   B. Identify why you believe the child is not performing this task.
   C. Suggest what the parent can do to help. (Hint: see sections on basic facts and place value in Chapter 9.)

Suggested Readings

Cansler, D. P., Martin, G. H., & Valand, M. C. *Working with families.* Winston- Salem: Kaplan Press, 1976.

Cross, L., & Goin, K. W. *Identifying handicapped children: A guide to casefinding, screening, diagnosis, assessment, and evaluation.* New York: Walker and Co., 1977.

Hobbs, N. *Issues in the classification of children.* San Francisco: Jossey-Bass, 1975.

Lovitt, T. C. *In spite of my resistance . . . I've learned from children.* Columbus, Ohio: Charles E. Merrill, 1977.

Pasanella, A. L., & Volkmor, C. B. *Coming back . . . or never leaving.* Columbus, Ohio: Charles E. Merrill, 1977.

Russell Sage Foundation. *Guidelines for the collection, maintenance and dissemination of pupil records.* Sterling Forest, N.Y.: Russell Sage Foundation, 1969.

Reference

Federal Register. Washington, D.C.: U.S. Government Printing Office, August 23, 1977.

Appendices

Appendices

Teacher's Review of Selected Mathematics Content

Axioms

Commutative Property
Addition: The order of addends does not affect the sum.

$$a + b = b + a$$

Multiplication: The order of factors does not affect the product.

$$a \times b = b \times a$$

Associative Property
Addition: The grouping of addends does not affect the sum.

$$(a + b) + c = a + (b + c)$$

Multiplication: The grouping of factors does not affect the product.

$$(a \times b) \times c = a \times (b \times c)$$

Distributive Property for Multiplication over Addition
For any three numbers, *a*, *b*, and *c*,

$$a (b + c) = ab + ac.$$

This axiom underlies the multiplication algorithm as shown below:

$$
\begin{array}{r}
23 \\
\times 4 \\
\hline
12 \\
80 \\
\hline
92
\end{array}
\qquad \text{or} \qquad
\begin{aligned}
4 \times 23 &= 4 \times (20 + 3) \\
&= (4 \times 20) + (4 \times 3) \\
&= 80 + 12 \\
&= 92
\end{aligned}
$$

Identity Element

Addition: When zero is added to any number, the sum is that number.

$$n + 0 = n$$

Multiplication: When any number is multiplied by 1, the product is that number.

$$1 \times n = n$$

Inverse

Addition: For each number, *n*, there is a number −*n* such that their sum is zero. If *n* is a negative number, then its additive inverse is positive.

$$n + (-n) = 0$$

Multiplication: For each number *n* ($n \neq 0$), there is a number 1/n such that their product is 1.

Closure Principle

Given a set of numbers and an operation on these numbers, if the answer (resulting number) is a member of the same set as the original numbers, the set is *closed* for that operation.

Addition: For all numbers *a* and *b*, if *a* and *b* are in a set *X*, then *a* + *b* is in set *X*.

Multiplication: If *a* and *b* are in set *X*, then *ab* is in set *X*.

Relations

A *relationship* is an association of some kind—for example, having something in common.

A *relation* in mathematics is a set of pairs of elements arranged according to an ordering rule, such as a given sequence.

An *ordered pair* consists of two elements that are in a certain sequence; for example, (2, 3) is different from the ordered pair (3, 2).

A *function* is a set of ordered pairs in which no two of the ordered pairs have the same first element. Thus, the set of ordered pairs $\{(1, 3), (2, 5), (3, 7)\}$ is not only a relation but also a function.

The relations "is less than" and "is greater than" are inverse relations.

The relations "is as tall as," "is as heavy as," "is parallel to," are their own inverses. They are called *symmetric relations*.

A *reflexive* relation is represented by "is equal to itself."

A *transitive* relation is represented by the following: "A is heavier than B; B is heavier than C; A is heavier than C"; "A is equal to B; B is equal to C; A is equal to C."

A relation that is symmetric, reflexive, and transitive is called an *equivalence relation*, such as "is the same age as."

Real Number System

Set of Natural Counting Numbers

$$\{1, 2, 3, 4, \ldots\}$$

Set of Whole Numbers

$$\{0, 1, 2, 3, 4, \ldots\}$$

Set of Integers

$$\{\ldots, -4, -3, -2, -1, 0, 1, 2, 3, 4, \ldots\}$$

Set of Rationals

$$\left\{
\begin{array}{l}
\ldots -4/1, \ -3/1, \ -2/1, \ -1/1, \ 0/1, \ 1/1, \ 1/2, \ 1/3, \ 1/4, \ldots \\
\ldots -4/2, \ -3/2, \ -2/2, \ -1/2, \ 0/2, \ 1/2, \ 2/2, \ 3/2, \ 4/2, \ldots \\
\ldots -4/3, \ -3/3, \ -2/3, \ -1/3, \ 0/3, \ 1/3, \ 2/3, \ 3/3, \ 4/3, \ldots
\end{array}
\right\}$$

Set of Irrationals

The set of numbers that may be expressed as an infinite nonrepeating decimal.

Prime and Composite Whole Numbers

Prime Number

A whole number greater than 1 that has no factors other than itself and 1. For example:

2, 3, 5, 7, 11.

Composite Number

A whole number c that can be expressed as a product $a \times b$, where both a and b are whole numbers other than 1 or 0. The numbers a and b are factors of c. A composite number has at least two factors other than 1 or 0, but may have more than two. For example:

$$30 = 2 \times 3 \times 5.$$

Fundamental Theorem of Arithmetic

Any whole number greater than 1 may be written as the product of a set of primes unique to that number. For example, the prime factors of 12

$$(3 \times 2 \times 2 \text{ or } 2 \times 3 \times 2)$$

are the same regardless of the sequence of the factorization.

Greatest Common Factor (GCF)

The GCF is the largest number that is a factor of each of two or more given numbers. For example, the GCF of the numbers 12 and 18 is 6.

Least Common Multiple (LCM)

The LCM is the smallest multiple that is common to two or more numbers. For example, the LCM of 4 and 6 is 12.

Decimal Fractions and Decimals

Decimal Fraction

A decimal fraction is equivalent to a common fraction whose denominator is a power of 10. The decimal fraction of 5/10 is written in the form .5.

Mixed Decimals

These are written in the form 48.24 and are composed of a whole number and a decimal fraction.

Finite Decimals (Terminating Decimals)

These contain a specific (finite) number of digits—for example, 4.7, .34.

Repeating Decimals (Infinite or Nonterminating)

These have the same digit or pattern of digits repeated. For example:

.666 or .285714285714.

Every finite decimal may be expressed as a repeating decimal by annexing zeros.

Nonrepeating Decimals (Infinite)

These represent irrational numbers. For example

$$\pi \ (3.14159 \ldots) \quad \text{or} \quad \sqrt{2} \ (1.41428 \ldots).$$

Scientific Notation

Allows large numbers to be expressed as a product of a mixed decimal and a power of 10. The units place contains a digit 1 to 9, as shown here:

$$468739 = 4.7 \times 10^5$$
$$31246 = 3.1 \times 10^4$$
$$1234 = 1.2 \times 10^3$$
$$28 = 2.8 \times 10^1$$
$$.431 = 4.3 \times 10^{-1}$$
$$.00093 = 9.3 \times 10^{-4}$$

Percents
Percents are decimal fractions equivalent to common fractions whose denominators are hundredths. The % sign takes the place of both the 100 as denominator and the decimal dot.

$$25/100 = .25 = 25\%$$

Transformational Geometry and Topological Concepts

Rigid Transformations
Translation (slide):   May be described as "moving along a line," as in moving a toy car along a path.
Rotation (turn):   Exemplified by "turning through an angle," as in the hands on a clockface.
Reflection (flip):   Exemplified by mirror images—d and b, p and q.

Projective Transformations
These involve changes in visual perception, such as projecting a shadow.

Topological Equivalence
The elements in topology include geometric figures such as points, lines, curves, surfaces, triangles, circles, and spheres. Topological equivalence means that two geometric figures can be twisted or stretched into each other's shapes.

Four basic topological relations:

- Proximity—closeness of things
- Separateness—configurations of shapes
- Enclosure—boundaries of bodies and objects
- Order—orders of things

Network is an aspect of topology that involves arcs and vertices. A *path* in a network is a set of one or more arcs and vertices that can be traced without lifting the pencil and without retracing an arc (line) of the network. The diagram on page 236 pictures a network that is not traversable (a) and one that is traversable (b).

a

b

Probability

Sometimes two events are equally likely. In other cases, certain events are likely to occur more frequently than others. The following questions relate to ideas underlying probability:

- Are all outcomes equally likely?
- If two or more events are involved, are they mutually exclusive? For example, when drawing only one card from a deck, you can draw an ace and a heart simultaneously (ace of hearts), but you cannot draw a heart and a diamond—these two events cannot occur simultaneously on a single draw.
- If two or more events are involved, are they independent? For example, if you (a) draw a card from a deck, then (b) replace it and shuffle the deck, and then (c) draw another card, the two drawings are independent.

Probability Examples and Solutions

Find the probability (P) of getting heads when flipping a dime. There are two possible outcomes, heads or tails, so

P (heads) = 1/2.

Find the probability of drawing an ace from a shuffled deck of cards. There are four aces and 52 possible outcomes.

P (ace) = 4/52 = 1/13.

State the probability of rolling a die and obtaining an even number. The numbers 1, 2, 3, 4, 5, 6 are represented by dots on the six different faces of the die. Of these six outcomes, three are even numbers: 2, 4, 6.

P (even) = 3/6 = 1/2.

State the P of rolling a die and obtaining a 14. The P of an impossible event is 0, therefore

P (14) = 0/6 = 0.

Reisman-Kauffman Interview Checklist

B

| Appearance | Characteristics | | Examples |
	Positive	Negative	
Dress	☐appropriate	☐inappropriate	
Facial expression	☐alert	☐placid	
	☐attentive	☐vacant	
	☐happy	☐sad	
Physical health	☐good	☐poor	
	☐appropriate weight	☐too fat/too thin	
Movement			
Observed	☐quick	☐slow	
	☐well-coordinated	☐clumsy	
	☐calm	☐restless	
		☐perseveration of	scratching, rubbing, mannerisms with hands
Inferred	☐purposeful	☐no purpose	
	☐meaningful	☐no meaning	
	☐moves freely	☐inhibited	
Emotional			
Expression	☐pleasant	☐sulky	
	☐relaxed	☐anxious	
	☐trusting	☐suspicious	
Mood	☐steady, constant	☐fluctuates, labile	inconsistency between child's appearance and how he describes his characteristic mood
	☐normal reaction	☐overreaction	shows greater emotional response than expected in relationship to his age and the situation
Special symptoms		☐delusions	
		☐hallucinations	
		☐compulsive	
		☐misinterprets environmental cues	another child knocks a book off his desk and he takes this as a deliberate act of aggression

237

Orientation	Characteristics		Examples
	Positive	*Negative*	
Spatial	☐knows where he/ she is	☐confused	
	☐knows how to get from one place to another	☐confused	
Temporal	☐knows time of day	☐confused	
	☐on time usually	☐late usually	
Verbal			
Talk	☐a lot	☐little	
	☐spontaneously	☐answers only	To whom does he talk spontaneously? Does he wait to be asked questions?
	☐coherent conversation	☐confused ☐sudden silences	
	☐stays on topic for appropriate time	☐keeps changing the subject ☐recurrent theme	
	☐appropriate use of tense	☐inappropriate tense	
	☐appropriate level of vocabulary	☐limited vocabulary ☐incorrect use of inflectional endings ☐incorrect use of prepositions	
	☐appropriate level of syntax	☐limited syntax	
Interpersonal Relationships			
Peers	☐seeks out peers	☐isolate	
	☐leader	☐follower ☐bossy	
	☐friends are of appropriate age	☐friends are of inappropriate age	picks friends much older or much younger
	☐participates in group	☐withdraws from group ☐rejected by group	
Adults	☐friendly	☐fearful ☐clinging ☐teacher's pet	

Orientation	Characteristics Positive	Negative	Examples
Interaction in General			
Rules	☐accepts them reasonably	☐overconcerned with	shows passive resistance, subversive behavior, openly rebellious
Instructional program	☐joins in enthusiastically	☐just goes along ☐disruptive	
	☐shows initiative ☐suggests new ideas	☐needs direction	
	☐handles frustration well	☐handles frustration poorly	losing a game, poor grade
	☐concentrates well	☐distractible during lesson	mathematics, spelling, lecture
	☐creative responses ☐interested	☐preoccupied	
	☐fluency of ideas ☐flexible	☐offers few ideas ☐rigid	
	☐original	☐mundane	
	☐tolerates ambiguity	☐needs structure	
	☐resists premature closure	☐needs immediate closure	

Instructional Strategies (IS) for Specific Learning Disabilities in Mathematics (SLDM) Checklist

C

SLDM	Related Component of the Developmental Mathematics Curriculum	IS
Reasoning ☐Does not appropriately sequence occurrences or objects	5.1.0 Sequencing; 5.3.0 Order; 6.6.0 Seriation; 9.4.0 Computation; 9.5.0 Seriation Extended	5.1.1; 5.1.2; 5.1.3; 5.1.9; 10.1.1
☐Displays logical errors in computing algorithms	9.4.0 Computation	10.1.1; 10.1.2; 10.1.3
☐Inability to make choices and decisions	6.7.0 Judging and Estimating; 9.4.0 Computation	5.1.5; 5.1.6; 5.1.10; 10.1.1
☐Does not connect cause–effect relationships	8.1.0 Transformations; 9.7.0 Cause–Effect Relationships	5.1.3; 5.1.9; 5.1.10; 5.3.2
☐Does not make appropriate inferences from data and draws inappropriate conclusions	6.7.0 Judging and Estimating; 9.7.0 Cause–Effect	5.1.9; 5.1.10; 5.3.2
☐Reverses word meanings, such as antonyms	8.4.0 Mathematics as a Language of Relationships	5.2.1; 5.3.1; 5.3.2; 5.4.1
☐Uses cues incorrectly	9.4.0 Computation	5.1.3; 5.1.4; 5.1.6; 5.3.1; 5.3.2; 5.4.1; 5.4.3; 5.4.6; 10.1.1
☐Incorrect use of syntax in language	8.4.0 Mathematical Language	8.4.4; 8.4.5; 8.4.8
☐Displays receptive language disabilities	8.4.0 Mathematical Language	8.4.2; 8.4.6

SLDM	Related Component	IS
Problem Solving		
☐Does not organize a task to facilitate its completion	6.7.0 Judging; 9.4.0 Computation	8.1.4; 10.1.4
☐Does not employ problem-solving strategies; often uses random, trial-and-error approach to tasks	5.1.0 Sequencing; 6.6.0 Seriation; 9.5.0 Seriation Extended	5.1.6; 5.1.7; 5.1.8; 5.1.11; 5.2.4; 5.3.1 5.3.2; 5.4.1
☐Does not complete work on time; seems to always be in a "duel" with time	9.4.0 Computation All tasks	10.1.4; 11.3.2
Orientation		
☐Does not indicate knowledge of spatial relationships (direction, location, laterality, topological relationships); disoriented spatially	5.3.0 Topological Relation ships; 9.4.0 Computation	5.1.3; 5.3.2; 9.1.1
☐Does not show awareness of temporal relationships (e.g., always late)	9.14.0 Applied Mathematics —Measurement	10.1.4
☐Reversals and mirror images displayed in writing digits, place value notation, alphabet, words	5.4.0 Arbitrary Associations; 8.4.0 Mathematical Language; 9.4.0 Computation	5.1.3; 5.1.4; 5.1.6; 5.3.1; 5.3.2; 5.4.1 5.4.2; 5.4.3; 5.4.6 5.4.8
Motor Performance		
☐Clumsy for age	9.4.0 Computation	10.1.2; 10.4.1
☐Excess tongue movement and/or salivation while writing mathematics computations	All topics	10.1.3
☐Excess body movement, especially as mathematics becomes more difficult	All topics	10.1.3
☐Writing is messy and results in computation errors	9.4.0 Computation	10.1.2
☐Immature eye–hand coordination	9.4.0 Computation	10.1.1; 10.1.2
☐Good gross-motor but poor fine-motor coordination	9.4.0 Computation	10.1.2
☐Displays expressive language disabilities	8.4.0 Mathematics as a Language of Relationships; 9.4.0 Computation	10.1.2

241

SLDM	Related Component	IS
Attention		
☐ Distractible	All topics	5.1.1; 5.3.1; 10.1.3; 11.1.4
☐ Attention flips in and out	9.4.0 Computation	11.1.5; 12.1.3
Perception		
☐ Displays visual perception problems	5.1.0 Sequencing; 5.3.0 Topological Relationships; 5.4.0 Arbitrary Associations;	5.1.3; 5.1.4; 5.1.5; 5.1.6; 5.2.1; 5.2.2; 5.2.4; 5.3.1; 5.3.2;
☐ Displays auditory perception problems	6.4.0 Sorting; 6.5.0 Many-to-one and one-to-many matching; 6.6.0 Seriation; 6.7.0 Judging and Estimating; 7.1.0 Concepts; 8.1.0 Transformations; 8.3.0 Equivalence Relations; 8.4.0 Mathematical Language	5.4.4; 5.4.5; 5.4.6; 5.4.7; 5.4.8
Affect		
☐ Is often ambivalent regarding self-concept	All topics	12.2.1; 12.3.1
☐ Inappropriate level of anxiety in relationship to task difficulty	9.4.0 Computation	10.1.1; 10.1.3; 10.1.4; 11.3.3

Developmental Mathematics Curriculum— Screening Checklist

D Indicate what the child *can* do by adding check marks.

Early Childhood Education Range: Grades K–4

I. *Basic Relationships*
  A. Prenumber Relationships
    1. Succession (spatial or temporal activities) (5.1.0)
      _____ a. Can follow directions, e.g., "Put on your sock, then your shoe."
      _____ b. Can walk in normal manner.
      _____ c. Climbs stairs in succession, with one foot per step.
      _____ d. Performs self-help skills in proper succession; for example, swallows before putting more food in mouth, tracks objects with eyes.
      _____ e. Can state succession of: morning–night.
      _____ f. Can state succession of: yesterday–today–tomorrow.
    2. Sequencing (succession with a pattern) (5.1.0)
      _____ a. Can count from 1 to 5 by rote.
      _____ b. Can count by ones from 1 to 10 by rote:____ to 15,____ to 20,____ to 50,____ to 100.
      _____ c. Completes dot pictures in numbered sequence from 1 to 10,____ 1–25.
      _____ d. Repeats digit sequences.
      _____ e. Repeats chain of objects; e.g., "doll, car, baby, dog."
    3. Matching (5.2.0)
      _____ a. Can match sets in one-to-one correspondence:____ through groups of 5,____through 10.
      _____ b. Can enumerate objects in a set:____5,____10.
      _____ c. Can match shapes on simple form boards.
  B. Topological Relationships (5.3.0)
    1. Proximity
      _____ a. Can place an object close to a given object.
      _____ b. Can draw a figure close to a given figure.
      _____ c. Can stand close to a given person.
    2. Enclosure
      _____ a. Can place an object inside a box.
      _____ b. Can stand within a boundary.
      _____ c. Can mark X inside a circle.

243

3. Separateness
    _____ a. Can select an open figure, such as a cut rubber band from those uncut.
    _____ b. Can take apart a puzzle.

4. Order
    _____ a. Can order:___events,___lengths,___size,___number.
    _____ b. Writes in missing numerals in a sequence:___1–5, ___1–10.
    _____ c. Names ordinal positions:___first–last,___first, second, third, fourth, fifth.
    _____ d. Matches ordinal positions to numbers 1, 2, 3, 4, 5.

C. Arbitrary Associations (5.4.0)
1. Words
    _____ a. Can name numbers "zero" through "ten."
    _____ b. Can say number names "zero" through "hundred."
    _____ c. Reads number words to *ten*.

2. Symbols
    _____ a. Reads and writes number symbols___0–5, ___0–9,___ 0–10, ___ 0–25, ___ 0–100.
    _____ b. States age, address, phone number.
    _____ c. Writes age, address, phone number.
    _____ d. Uses adapted symbol system:___rebus,___Bliss,___Braille,___Fitzgerald Key.

II. *Lower-Level Generalizations*
A. Duality (6.1.0)
1. Symmetry
    _____ a. Can point out symmetry in own body.
    _____ b. Can draw symmetry in animals.

2. Same–Different
    _____ a. Can state similarities among objects.
    _____ b. Can state differences among objects.
    _____ c. Can select similar patterns.

B. Set Relationships
1. Equivalence of Sets (6.2.0)
    _____ a. Selects equivalent sets.
    _____ b. Can draw a set equivalent to a given set.
    _____ c. Can create a set equivalent to a given set, but using a different class of objects.
    _____ d. Correctly uses = sign.

2. Greater Than–Less Than Matching (6.3.0)
    _____ a. Selects a set larger than a given set.
    _____ b. Selects a set smaller than a given set.
    _____ c. Creates a set larger than a given set.
    _____ d. Creates a set smaller than a given set.
    _____ e. Creates a set "*one* more than" a given set.
    _____ f. Creates a set "*one* less than" a given set.
    _____ g. Draws a set:___greater than,___less than a given set.
    _____ h. Correctly uses < , > signs.

3. Sorting (6.4.0)
    _____ a. Can sort by:___shape,___size,___length,___color, ___texture, ___ function.

4. Many-to-one/one-to-many matching (6.5.0)
    _____ a. Can match five fingers to one hand.
    _____ b. Uses tally in scoring games, e.g. ///// or a notch on a stick to represent five.
    _____ c. Describes matching on a picture graph; for example, many children belonging to the same school or family.

C. Seriation (6.6.0)
  1. Numerical
    _____ a. Can fill in missing numbers in a series.
    _____ b. Can fill in a number line.
    _____ c. Can count by:____2's,____5's,____10's.
    _____ d. Can arrange dominoes by number in order.
  2. Geometric
    _____ a. Can arrange figures by number of sides.
    _____ b. Can arrange figures by pattern.
    _____ c. Can arrange given series by size or length.

D. Judging and Estimating (6.7.0)
  1. Binary judgments
    _____ a. Can tell if an object is *red/not red*.
    _____ b. Can state if picture shows *cold/not cold* people.
  2. Multiple judgments
    _____ a. Can select amount of food for dinner.
    _____ b. Can make simple inferences based on judgments and/or estimations.

III. *Concepts* (7.1.0)
A. Number
  1. Cardinality
    _____ a. Points to empty set.
    _____ b. Points to sets of:____1,____2,____3,____4,____5,____6–10.
    _____ c. Enumerates sets from:____0–5,____0–10.
    _____ d. Names cardinality of sets with____0–5 objects,____0–10 objects.
    _____ e. Selects fractional part of object—for example, one-half.

B. Geometry
  1. Shapes
    _____ a. Names geometric shapes such as circle, square, rectangle.
    _____ b. Traces geometric shapes.
    _____ c. Copies geometric shapes.

C. Measurement
  1. Size and amount
    _____ a. Uses correct measuring device—thermometer for temperature, ruler for length.
    _____ b. Can select objects according to the following size and amount labels:____ big–little, ____ long–short, ____ tall–short,____fat–thin,____old–young,____far–near.
  2. Time
    _____ a. Can state that clocks are used to tell time.
    _____ b. Can tell time of day; e.g., morning or night for: ____get up in morning,____go to school,____eat lunch,____eat dinner.
    _____ c. Can tell month for holidays (Thanksgiving, Christmas).

3. Money

       _____ a. Can point to the following coins when named:____ penny,____nickel,____dime.

       _____ b. Can state the following values: ____ 5 pennies = 1 nickel, ____ 5 nickels = 1 quarter, ____ 10 pennies = 1 dime, ____ 25 pennies = 1 quarter, ____ 2 nickels = 1 dime, ____ 2 dimes and 1 nickel = 1 quarter, ____ 2 quarters = 1 half-dollar, ____ 4 quarters = 1 dollar, ____2 half-dollars = 1 dollar, ____ 20 nickels = 1 dollar, ____ 10 dimes = 1 dollar, ____ 100 pennies = 1 dollar.

       _____ c. Pays for school item (e.g., lunch).

       _____ d. Tells difference between buying and selling.

       _____ e. Tells difference between spending and saving.

       _____ f. Uses cents sign (¢).

       _____ g. Writes monetary value in form 5¢.

       _____ h. Counts change up to $1.00.

       _____ i. Counts paper money up to $5.00.

       _____ j. Can make change up to $1.00.

IV. *Higher-Level Relationships*

  A. Transformations (8.1.0)

    1. Conservation

       _____ a. Conserves number.

       _____ b. Conserves mass.

       _____ c. Conserves time.

       _____ d. Conserves length.

       _____ e. Conserves discontinuous quantity (liquid).

       _____ f. Conserves volume.

    2. Slides/Turns/Flips

       _____ a. Can state that an object has only been moved in space but has no other changes.

       _____ b. States that a rotated object changes only its position.

       _____ c. Selects a picture of a mirror image of an object.

  B. Set Operations and Relationships (8.2.0)

    1. Union and Intersection

       _____ a. Uses union of sets as a model for addition.

       _____ b. Can tell what objects belong to two or more sets.

       _____ c. Can use $\cup$ and $\cap$ signs.

    2. Subset—class inclusion

       _____ a. Can represent class inclusion with concrete objects.

       _____ b. Can show subset relationship with pictures or diagrams.

       _____ c. Can use symbols to mean subset.

       _____ d. Can select disjoint sets.

  C. Number Operations and Relationships (8.3.0)

    1. Binary operations (+, ×, −, ÷)

       _____ a. Can write correct operation sign to make statement true such as

$$3 \square 2 = 5, 3 \square 2 = 1, 6 \square 2 = 12, 6 \square 2 = 3.$$

       _____ b. Can tell what operation to use in a given situation, e.g., "I want to find the whole," "I want to find what is left," "I want to find the number of equal-size parts."

2. Inequalities
　　———————— a. Selects the smaller number.
　　———————— b. Selects the larger number.
　　———————— c. Uses the greater-than and less-than signs in number rela-
　　　　　　　　　tionships (e.g., 4 □ 7).
　　———————— d. Compares objects according to terms shown earlier
　　　　　　　　　under III *Concepts*, C1b (big–little, old–young, etc.).
　　———————— e. Uses greater-than/less-than symbols with numerals
　　　　　　　　　0–25.
3. Equivalence Relations
　　———————— a. Selects situations showing reflexive relations, such as
　　　　　　　　　"the same as."
　　———————— b. Selects situations showing symmetric relations, such as
　　　　　　　　　"is as tall as."
　　———————— c. Selects situations showing transitive relations, such as
　　　　　　　　　"if John is taller than Mike, and Mike is taller than Bob,
　　　　　　　　　then John is taller than Bob."
　　———————— d. Selects concrete or pictorial examples of equivalent
　　　　　　　　　fractions (1/2 and 2/4).

D. Mathematical Language (8.4.0)
1. Inflectional Endings
　　———————— a. Selects the longer, shorter, heavier, lighter of two
　　　　　　　　　objects.
　　———————— b. Selects the longest, shortest, heaviest, lightest of three
　　　　　　　　　or more objects.
　　———————— c. Selects the number of objects indicated, such as sin-
　　　　　　　　　gular–plural.
2. Prepositions
　　———————— a. Follows directions involving: ———on, ———under, ———
　　　　　　　　　above, ———below, on top of, ———underneath, ———
　　　　　　　　　beneath,———in front of,———behind,———over,———up,
　　　　　　　　　———to the left of,———to the right of,———in the
　　　　　　　　　middle of, ———around.
3. Temporal
　　———————— a. Uses tense correctly:———present,———past,———future.
　　———————— b. Uses *before–after* correctly.
4. Fractions
　　———————— a. Can label parts of a whole:———one-half,———one-fourth,
　　　　　　　　　———one-third,———two-fourths,———one-tenth,———one-
　　　　　　　　　hundredth.
　　———————— b. Can label fractional part of a group.

V. *Higher-Level Generalizations*
A. Basic Facts (9.1.0)
1. Addition
　　———————— a. Can compute sums to 9.
　　———————— b. Can compute sums to 18.
2. Subtraction
　　———————— a. Can compute with minuends to 9.
　　———————— b. Can compute with minuends to 18.
3. Multiplication
　　———————— a. Can compute with products to 25.
　　———————— b. Can compute with products to 81.

4. Division
_____ a. Computes with dividends to 25.
_____ b. Computes with dividends to 81.
5. Symbols
_____ a. Uses +, −, ×, ÷ signs correctly.

B. Axioms (9.2.0)
1. Addition
_____ a. Can select commutative property for addition

$$a + b = b + c.$$

_____ b. Can select associative property

$$(a + b) + c = a + (b + c).$$

_____ c. Can use the additive identity

$$n + 0 = n.$$

_____ d. Can use the additive inverse property

$$n - n = 0.$$

_____ e. Can select sets that are closed for addition (set of even numbers, set of whoe numbers, etc.).

2. Multiplication
_____ a. Can select commutative property

$$a \times b = b \times a.$$

_____ b. Can select associative property

$$(a \times b) \times c = a \times (b \times c).$$

_____ c. Can use the multiplicative identity

$$n \times 1 = n.$$

_____ d. Can use the multiplicative inverse

$$n \times 1/n = n/n = 1.$$

_____ e. Can select sets that are closed for multiplication (set of even whole numbers, set of odd whole numbers).
_____ f. Can use the distributive property of multiplication over addition for whole numbers

$$a \times (b + c) = (a \times b) + (a \times c)$$

3. Subtraction
_____ a. States that subtraction is not commutative.
_____ b. States that subtraction is not associative.

_____ c. Can use the right identity property $(n - 0 = n)$.
_____ d. States that the set of whole numbers is not closed for subtraction.

4. Division

_____ a. States that division is not computative.
_____ b. States that division is not associative.
_____ c. Can use right identity property $(n \div 1 = n)$.
_____ d. States that the set of whole numbers (and integers) is not closed for division.
_____ e. Can use right distributive property $[(a + b) \div c = (a \div c) + (b \div c)$.

5. *Rationals and Axioms*

_____ a. Can apply axioms to set of rationals____in fraction form,____decimal form.
_____ b. Can use multiplicative identity in the form n/n to rename improper fractions to simplest form; e.g., proper fraction, mixed number.

C. Place Value (9.3.0)

1. Notation

_____ a. Can write numerals 0–9 when dictated.
_____ b. Can write numerals 10–20 when dictated, ____0–99.
_____ c. Can write numerals 0–100,____0–999 when dictated.
_____ d. Can read numerals 0–9, ____10–20, ____0–99, ____ 0–100, ____0–999.
_____ e. Can write two digit numerals in expanded form:

$$32 = (3 \times 10) + (2 \times 1) \text{ or 3 tens plus 2 units.}$$

_____ f. Can write three-digit numerals in expanded form.
_____ g. Can write two-digit numerals as follows:

$$32 = \underline{2} \text{ units, } \underline{3} \text{ tens,}$$

three-digit in this form, such as

$$432 = \underline{\quad} \text{units,} \underline{\quad} \text{hundreds,} \underline{\quad} \text{tens.}$$

_____ h. Can rename place values:

$$20 \text{ units} = 2 \text{ tens, } 43 \text{ units} = 4 \text{ tens, } 3 \text{ units.}$$

Middle School Education Range: Grades 4–8

2. Application

_____ a. Generalizes place-value notation to decimal places by selecting tenths, hundredths.
_____ b. Can write decimal fractions in expanded form:

$$3.25 = (3 \times 1) + (2 \times 1/10) + (5 \times 1/100).$$

_____ c. Can write decimals in vertical algorithm form.

_____ d. Arranges objects into sets of ten with and without extra objects up to nine.

_____ e. Shows relationship of ten and units that comprise the "teens."

_____ f. Generates place values in____base 5,____base 2,____ base 3 (option for a gifted student).

D. Computation (9.4.0)

1. Addition—Whole numbers, integers, rationals, decimals

_____ a. Adds two-digit plus two-digit, no renaming.

_____ b. Adds multidigit plus multidigit, no renaming.

_____ c. Adds two-digit plus two-digit with renaming.

_____ d. Adds multidigit plus multidigit with renaming.

_____ e. Adds integers:____ + plus + , – plus –, and + plus –.

_____ f. Adds rationals—fractions with like denominators.

_____ g. Adds fractions with unlike denominators.

_____ h. Adds mixed numbers.

_____ i. Adds decimals without renaming.

_____ j. Adds decimals with renaming.

2. Multiplication—Whole numbers, integers, rationals, decimals

_____ a. Multiplies two-digit by one-digit, no renaming.

_____ b. Multiplies two-digit by two-digit, no renaming.

_____ c. Multiplies three-digit by two-digit, no renaming.

_____ d. Multiplies two-digit by one-digit, with renaming.

_____ e. Multiplies two-digit by two-digit, with renaming.

_____ f. Multiplies multidigit by multidigit, with renaming.

_____ g. Multiplies integers:____ +× +, ____ –×–, ____ + × –.

_____ h. Multiplies rationals with like and unlike denominators.

_____ i. Multiplies rationals in form of mixed numeral.

_____ j. Multiplies decimals without renaming.

_____ k. Multiplies decimals with renaming.

3. Subtraction—Whole numbers, integers, rationals, decimals

_____ a. Subtracts two-digit from two-digit, no renaming.

_____ b. Subtracts multidigit from multidigit, no renaming.

_____ c. Subtracts two-digit from two-digit, with renaming.

_____ d. Subtracts multidigit from multidigit, with renaming.

_____ e. Subtracts with zero in minuend, with renaming.

_____ f. Subtracts integers with and without renaming.

_____ g. Subtracts fractions with like denominators, no renaming.

_____ h. Subtracts fractions with unlike denominators, no renaming.

_____ i. Subtracts fractions with renaming (such as mixed numbers).

_____ j. Subtracts decimals without renaming.

_____ k. Subtracts decimals with renaming.

4. Division—Whole numbers, integers, rationals, decimals

_____ a. Divides two-digit by one-digit, without remainder.

_____ b. Divides two-digit by one-digit, with remainder.

_____ c. Divides multidigit by two-digit with and without remainder.

———— d. Divides three-digit by one-digit, no renaming in dividend, e.g., $2\overline{)468}$.

———— e. Divides three-digit by one-digit, with renaming in dividend, e.g., $2\overline{)356}$.

———— f. Divides multidigit by two-digit with and without renaming in the dividend.

———— g. Divides integers: ——— $+ \div +$, ——— $- \div -$, ——— $+ \div -$, ———
$- \div +$.

———— h. Divides rationals in the form of proper fractions.

———— i. Divides mixed number by mixed number.

———— j. Translates fraction to decimal.

———— k. Divides decimal by whole number.

———— l. Divides whole number by decimal, terminating and
nonterminating quotients.

———— m. Divides decimal by decimal, terminating and nonterminating quotients.

High School Education Range: Including Voc Ed

E. Seriation Extended (9.5.0)

———— 1. Completes simple analogies.

———— 2. Completes series number patterns.

———— 3. Completes patterns of geometric figures.

F. Solving Word Problems (9.6.0)

———— 1. Solves present-tense problems: ——— $+$, ——— x, ——— $-$,
——— $\div$.

———— 2. Solves past-tense problems: ——— $+$, ——— x, ——— $-$, ——— $\div$.

———— 3. Solves future-tense problems: ——— $+$, ——— x, ——— $-$, ——— $\div$.

———— 4. Solves mixed-tense problems: ——— $+$, ——— x, ——— $-$, ——— $\div$.

———— 5. Solves conditional problems; for example:

If I want 6 dogs and now have 4, how many do I need?

G. Cause-Effect (9.7.0)

———— 1. Shows awareness of cause–effect, if . . . then situations.

———— 2. Shows awareness of negation situations.

H. Topological Equivalence (9.8.0)

———— 1. Identifies transformations that are topologically equivalent.

I. Projective Geometry (9.9.0)

———— 1. Uses projective transformations in drawings.

J. Nonmetric Geometry (9.10.0)

———— 1. Can point out geometric shapes in environment:
——— circles, ——— squares, ——— triangles, ——— rectangles.

———— 2. Calls geometric shapes by their geometric name.

———— 3. Can find geometric shapes embedded in puzzle.

K. Metric geometry (9.11.0)

———— 1. Can use nonstandard measures.

———— 2. Can use standard measures:———length,———area,———
volume,———mass.

L. Network Theory (9.12.0)
    ———— 1. Can tell if a network is traversable using Euler's procedure.
    ———— 2. Can complete networks by drawing arcs when given odd and even vertices.
    ———— 3. To find relationships among the number of spaces, vertices, and arcs in any network.

M. Probability, Statistics, Graphs (9.13.0)
    ———— 1. Can use graphs to represent and interpret data, such as ——block graphs, ——pictographs.
    ———— 2. Can find probabilities.
    ———— 3. Can find mean, median, mode.

Early Childhood Education (contd.)

N. Applied Mathematics
   1. Measurement (9.14.0)—Time/Calendar
    ———— a. Can show time to the hour.
    ———— b. Can read time to the hour.
    ———— c. Can *show* time to the minute.
    ———— d. Can *identify* time to the minute, by stating "so many minutes after the hour."
    ———— e. Can tell time by fractions:——half hour,——quarter hour.
    ———— f. Can count by 5-minute intervals.
    ———— g. States there are 60 minutes in an hour,——30 minutes in half-hour,——12 hours in half-day,——distinguishes A.M. and P.M.
    ———— h. Can read so many minutes before the next hour.
    ———— i. States the number of days in a week,——months in a year.
    ———— j. Tells names of days in a week.
    ———— k. Can read names of days,——months.
    ———— l. Uses the calendar to interpret date.
    ———— m. Tells his or her birthday by——month,——day,——year.

Middle School Education (contd.)

    ———— n. Interprets written notation of time, e.g. 3:00, 3:01, 3:30, 3:45, etc.
    ———— o. Writes time shown on a clock face.
    ———— p. States number of seconds in a minute.
    ———— q. Sets an alarm clock.
    ———— r. Constructs time schedules, such as for homework or TV.
    ———— s. Uses time schedules—bus, train, plane, subway.
    ———— t. Reads public transportation schedules.
    ———— u. Explains Daylight Saving Time.

High School Education (contd.)

        _____ v. Can use a time clock.

        _____ w. Can explain meaning of *time and a half, double time.*

        _____ x. Is on time for school, work.

        _____ y. Explains and identifies different time zones in the world and can compute equivalent time across zone.

        _____ z. Uses time in relationship to____speed,____distance, ____cooking.

2. Measurement—clothing size

        _____ a. Can interpret sizes of____clothes,____shoes, ____underwear.

3. Measurement—temperature.

        _____ a. Can read and interpret temperature on a thermometer.

        _____ b. Can use temperature in cooking.

4. Metric Measure

        _____ a. Can relate metric measures to decimal notation.

        _____ b. Can use metric measures in cooking; weights.

        _____ c. Can convert values from one measure to another— metric to nonmetric, metric to metric, nonmetric to nonmetric (inches to feet, gallon to quarts).

        _____ d. Tells approximate distance between two locations in either kilometers or miles.

        _____ e. Uses measure in basic shop work.

        _____ f. Uses legend on map to estimate kilometers or miles.

        _____ g. Tells how to use an odometer; a speedometer.

5. Economics (9.15.0)

        _____ a. States the approximate money value of objects, such as soda pop, crayons, movie ticket, postage stamps.

        _____ b. Recognizes paper money to $20.00.

        _____ c. Makes change up to $25.00.

        _____ d. Writes all money amounts in decimal form.

        _____ e. Uses ads in newspapers, on radio, TV, and on store front windows to compare prices of food, clothing, etc.

        _____ f. Keeps daily record of expenditures.

        _____ g. Can check sales tax on purchases computing percent.

        _____ h. Computes hourly wage.

        _____ i. Displays understanding of state and federal taxes.

        _____ j. Displays understanding of social security tax and purpose of tax.

        _____ k. Can estimate living expenses, such as rent, utilities.

        _____ l. Can estimate cost of various licenses—auto, driver's, business.

        _____ m. States basic bank interactions and their purpose: checking account, savings account, loan, interest on savings accounts, interest on loans.

        _____ n. Can interpret bank statement.

        _____ o. Can keep checking account balanced.

        _____ p. States pros and cons of installment buying.

Some Trouble Spots in the Primary Grades Mathematics Curriculum

E Following are components of the mathematics curriculum that is typical for primary grade children. As a result of task analysis, clinical interviews with children, and observation, the following trouble spots have been identified.

I. Counting Objects versus Counting on a Number Line

Counting a series of objects in a set leads to obtaining the cardinality of the set of objects. However, this does not correspond to counting on a number line (or a clock face, which is really a circular number line). When counting objects, the *objects* are enumerated, not the spaces between them. But on a number line or a clock face, the spaces between—represented by jumps or moves—are counted rather than the dots or the minute marks. This discrepancy is illustrated in Figure E-1.

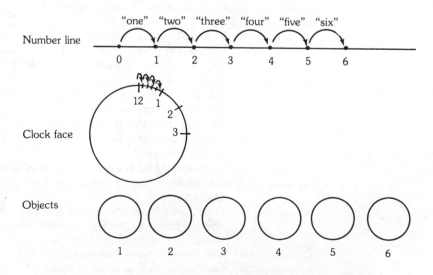

Figure E-1. *Discrepancy in Counting Moves and Objects*

II. Zero: Discrepancy between Physical Objects and Notation

When representing the number zero concretely or with pictures, no objects or pictures are used. However, *nothing is represented with something* when using digits—namely, the digit 0. Therefore, there is a discrepancy between the physical representation of zero and its digital representation. This discrepancy is apparent in place-value notation, when the digit zero must be used to represent none of a place. Zero is the only digit which has this discrepancy. All of the other digits, 1–9, may be represented by "something" in the real world. Students often are not aware of this discrepancy; for some, this causes misconceptions regarding place value.

III. Place-Value Placement in the Instructional Sequence

The numbers resulting from *ten ones* (10 × 1), *one ten* (1 × 10), *ten tens* (10 × 10), *ten hundreds* (10 × 100), and so on are products. A product is the result of performing the binary operation of multiplication on two numbers; the product is the answer to a multiplication problem. Place value is a product. Yet, when is place value taught? And when is multiplication taught? Place value is typically a first grade topic, while multiplication is introduced at the end of second grade and most usually in grade three. Unless the prerequisite of multiplication is taught either prior to or simultaneously with place value, only a rote learning of the place-value names occurs.

IV. Recording Numerals in Sequence ≠ Bundling by Tens

Writing numbers greater than nine in sequence is not the same as making a ten-for-one exchange. The ten-for-one exchange is a multiplication relation-ship, while counting in sequence is additive in character.

Children count aloud before they write numerals. When a child says "nine" and then "ten" in the early stages, he uses "ten," "eleven," and so on as the same type of label as "nine." Nine triggers "ten" in the *add-one* sequence, just as "eight" was the stimulus for "nine." The same thought processes are involved as the child begins to record this sequence with numerals. He may be aware that two digits are used to record ten, but he probably has learned this by rote with no thought to the notational principle of place value. The sequence model focuses on only two consecutive numbers at a time; it involves a stimulus-response series.

The bridge from writing 9 to 10 involves the idea of "nine and one more is ten." This is an activity prerequisite to bundling groups of ten, which is a many-to-one multiplication relationship requiring attention to the total array of numbers from 1 to 10 simultaneously. Bundling groups of 10 is an *exchange* model rather than a sequence model: ten units are exchanged for one *ten*. The exchange model requires that *ten* objects be counted and then a move occurs; the sequence model involves a change to a new place value position after counting *nine* objects. (See ISs 9.31-9.3.4, pp. 154-159.)

V. Teach Time to the Minute First

Children learn to count by *ones* first, not last. Therefore, in a time-telling sequence that includes counting by fives (such as "five minutes after the hour") or by fractions (such as "a quarter after the hour"), children should learn to tell time to the precision of a minute in *grade one*, rather than the end of grade two or in grade three. The traditional sequence introduces the more difficult parts of the task too early.

VI. Undo ≠ Inverse ≠ Reversibility of Thought

Undoing, taking apart, or separating a set into two or more of its component parts does not insure that one sees the relationship between inverse operations. Inverse operations in thought are consecutive actions—not reversed thought.

VII. Sequence of Instruction for the Binary Operations

Simple expressions of multiplication are found in a young child's environment:

> "I have three pieces of gum." (3 × 1)
> "I have four sisters." (4 × 1)
> "I ate two hot dogs." (2 × 1)
> "I have two bags of marbles and there's three in a bag." (implying three in *each* bag) (2 × 3)

The child's language expresses multiplication long before the operation is pointed out formally. As noted previously, being aware of this idea is a necessary prerequisite to understanding place value. The developmental nature of the mathematics curriculum is violated when multiplication is not even introduced until two years after topics based upon it are taught.

256

VIII. Discrepancy between Writing and Reading the Vertical Multiplication Algorithm

This discrepancy has been pointed out previously by Reisman (1977). The algorithm is usually written from top to bottom. For the example "three times forty-two" the progression is usually as follows:

Step 1	Step 2	Step 3	Step 4
42	42	42	42
	3	×3	×3

The completed algorithm is read in an upward direction, stating the multiplier first and then the multiplicand. Although the product is the same, this discrepancy is confusing when one wishes to distinguish between "three forty-twos" and "forty-two threes." Sometimes merely pointing out the discrepancy is sufficient for helping a confused student. For a student with learning problems it helps to suggest writing the algorithm in the same direction as subvocalizing or stating the problem—"forty-two, three times" for the sequence above or the following sequence for "three times forty-two":

Step 1	Step 2	Step 3	Step 4
3		42	42
	×3	×3	×3

IX. Discrepancy between Place-Value Notation and Use of a Number Line

Place values increase in a *right-to-left* direction; values on a number line increase in a *left-to-right* sequence. This is confusing to the student who is engaged in initial learning experiences involving both, as shown below:

hundreds	tens	units	tenths	hundredths
100 $>$	10 $>$	1 ...		

	zero	one	two	three . . . ten . . .
. . .-3 $<$ -2 $<$ -1 $<$	0 $<$	1 $<$	2 $<$	3 . . .

Also confusing is the function of the units place value as the fulcrum of our decimal notation system, in comparison with zero as the fulcrum for negative and positive integers.

Perhaps using a vertical number line would help to alleviate some of the confusion. Children have early experiences with vertical number lines— thermometers for showing temperature and graphs for recording measurement. Using the y–axis instead of the x–axis would allow the same manipulations (such as counting, computing, and sequencing) but would not interfere with learning place-value notation.

Case Studies and IEPMs

Case Study: **Lenny**

Referral Form:

NAME Lenny Jones   AGE 3 years 8 months
SCHOOL Bunny Hop PreSchool   GRADE Preschool   SEX M
PARENTS/GUARDIAN Mr. & Mrs. Harry Jones
ADDRESS Sunny Way Trailer Park   PHONE None
REFERRED BY Ms. Long   POSITION Preschool teacher
REASON FOR REFERRAL Evaluation by multidisciplinary team, medical consultation

PROBLEM AREAS

Acculturation and language: Normal for age
Health: Bed wetting, not toilet-trained, normal physical development
Hearing: appears normal
Vision: OK for age
Speech: OK for age
Perceptual-motor performance: OK for age
Academic achievement: performs tasks for age, can identify most primary colors, scribbles with crayons, seems to "catch on" at normal rate for age
Social-emotional behavior: displays extremes in emotion (mood swings from anger to silliness to pleasantness), hits other children
Parental involvement: Mother brings him to school every morning by bus. She tells of temper tantrums on the bus, but these are not apparent at school. Each morning mother hugs Lenny and says "I'll see you later." Father usually picks Lenny up by car in the afternoon—he often smells of alcohol.
Other comments: Lenny's response to behavior modification techniques has been minimal. Mother mentioned that she was taking older brother for an examination. We suggested that Lenny be examined also.

SUMMARY OF NEUROLOGICAL REPORT

Lenny was a three-year-old boy brought by his mother along with his brother, Carlton, who had some behavior problems. The mother stated that Lenny's problems were similar to those of his brother, but were of a milder nature. She said Lenny had temper tantrums and that when she would take him to stores he often would throw himself on the floor with a temper tantrum when he did not get what he wanted. She stated that he was not toilet-trained and soiled and wet his pants during the day as well as at night. The mother said he knew some colors; he scribbled; he could put on his sneakers, but not his socks or any other clothes. He could use a spoon and fork. He spoke in sentences. The mother stated that Lenny did not want to sleep in his crib but instead went to his sister's room and slept with her. He ate fairly well, had no allergies, and had had all his immunizations.

The pregnancy was uncomplicated. He was born in breech position approximately two weeks after the expected date. He was normal during the infancy. He walked and talked on time. He had not had any major surgery or trauma.

His head circumference was 51.5 cm, which was normal for him. He had no birthmarks or midline defects. His walking was normal for his age. He could throw a ball fairly well, and when the ball was rolled along the floor he went right to it without any hesitation. He scribbled with a pen and attempted to draw a circle. He built a tower of approximately eight blocks without problem. When the blocks were numbered or coded by color, he put them together by pattern.

In conclusion, this boy's physical and intellectual development were quite adequate for his age. He appeared to interact appropriately for his age. He was not toilet-trained, and this was a family trait. (The mother stated that his older brother was not trained until the age of 3, the mother was a bed wetter until the age of 5, and other family members had had problems with enuresis until the age of 16.)

Regarding the behavior problems, the home situation was not optimal due to the father's alcoholism and the influence of the mother's sisters. (Several of the mother's sisters, who visited often, had significant marital problems and earlier had been discipline problems, having to be placed in foster homes.) It was difficult to evaluate Lenny's hyperactivity, since he had been waiting for two hours during his brother's examination. He was somewhat cranky, but this behavior was not abnormal for the situation. He may well have been following the pattern of his older brother and aunts. He was too young for drug therapy. Perhaps behavior modification utilizing a reinforcement that he liked would be indicated—for example, color with crayons, string block patterns. Also, family counseling might be suggested in which appropriate role models might be learned by the parents and/or sisters.

TEACHER'S SUMMARY OF INTERVIEW CHECKLIST (APPENDIX B)

Appearance: appropriate dress, facial expression is usually alert, attentive, happy; physical health is good, weight is appropriate

Movement: observed—quick, well-coordinated, restless; inferred—purposeful, meaningful, moves freely

Emotional: expression is pleasant, anxious, trusting; mood is labile, fluctuates, overreactive on occasion

Orientation: spatial—knows where he is, knows how to get from one place to another; temporal—usually on time but confused as to time of day

Verbal: talks a lot, spontaneously; conversation is coherent, stays on topic for appropriate time; appropriate level of vocabulary

Interpersonal relationships: seeks out peers, friends of same age, friendly to adults

Interactions in general: rejects rules rebelliously, disruptive to instructional program, shows initiative, handles frustration poorly, distractible during lessons except artwork, gives some creative responses, fluency of ideas, original, resists premature closure

TEACHER'S SUMMARY OF IS-SLDM CHECKLIST (APPENDIX C)

Reasoning: *strengths*—can sequence objects, can make simple choices and decisions; *weaknesses*—cannot sequence occurrences, does not connect cause–effect relationships, does not make inferences, reverses word meanings (big–little), prepositions
Problem solving: *strength*—completes work on time
Orientation: *strength*—shows knowledge of spatial relationships; *weakness*—lacks awareness of temporal relationships
Motor performance: *strength*—good coordination; *weakness*—excess body movements most of the time
Attention: *weakness*—distractible
Perception: seems OK
Affect: *weakness*—rebellious, temper tantrums

TEACHER'S SUMMARY OF DEVELOPMENTAL MATHEMATICS CURRICULUM CHECKLIST (APPENDIX D)

I. *Basic Relationships*: *strengths*—succession (walks in normal manner, performs self-help skills, can state succession of morning–night); sequencing (repeats chain of objects); matching (can match shapes on simple form boards); proximity, enclosure, separateness; can say number names to "three." Reads digits 1, 2, 3; *weaknesses*—cannot follow directions, cannot state succession of yesterday, today, tomorrow; cannot repeat digit sequences; cannot match sets through 5 in 1–1 correspondence; order
II. *Lower-Level Generalizations*: *strengths*—symmetry, sorting by shape, color; binary judgments; *weaknesses*—same–different, sorting by size, length, texture, function; multiple judgments

---

IEPM for Lenny

Child's name: Lenny Jones   *Subject Area*: Mathematics
Level of Performance: not toilet-trained; can identify primary colors; cognitive and physical development normal; social-emotional behavior: tantrums, labile behavior, shows hyperactivity during most of preschool activities; good spatial relationships; can sequence objects; make simple choices and decisions; cannot sequence events; reverses meaning of big–little and prepositions; performs most of the basic relationship tasks; able to sort by shape and color; cannot sort by size, length, texture, function; cannot make multiple judgments

Prioritized Long-Term Goals:

I. To develop and refine basic relationships and lower-level generalizations
II. To improve interpersonal interactions

Short-Term Objectives	IS	Special Educational And/or Support Services	Persons Responsible	Length of Time
Long-Term Goal I				
To sequence events	5.2.4		Preschool teacher	3 months
To sort by function, texture	5.1.1; 5.1.3		"	1 month
Long-Term Goal II				
To make friends	12.1.5, 12.2.1, 12.3.1	Counseling services, parent counseling and training, social work services	Related services personnel	12 months

Summary: Lenny

The IEPM committee comprised Mrs. Long, Lenny's preschool teacher; the preschool director; the parents; and the psychoeducational consultant for the preschool, who was able to relate the medical information, having been in contact with both the pediatric neurologist and the family's physician.

Since Lenny was so young, the parents stated that they realized that the home situation must have caused a lot of the boy's behavior problems. They told about the IEPM committee meeting with their older son's school. They stated that they had decided to enter a program for family counseling services.

The committee agreed that Lenny had shown many strengths in basic relationships and generalizations in mathematics, apparent from the checklists. They agreed that he was developing mathematics competencies as expected and that work in this area would continue with the instructional strategies indicated on the IEPM.

Case Study: **Sali**

Referral Form:

NAME   Sali Pine   AGE   10 years 4 months
SCHOOL   Berry Road Middle School   GRADE   5   SEX   F
PARENTS/GUARDIAN   Dr. and Mrs. Roger Pine
ADDRESS   14 Berry Road   PHONE   WA 3-3159
REFERRED BY   Mr. Cohen   POSITION   5th grade teacher
REASON FOR REFERRAL   Evaluation by multidisciplinary team, learning
                                    disabilities consultation

PROBLEM AREAS

Acculturation and language: family is multilingual (French, Russian and Hebrew in addition to English), lived in Israel when Sali was 3–6 years old.
Health: excellent but a little overweight

Hearing: appears normal

Vision: wears glasses to correct for nearsightedness

Speech: stammers slightly under stress in class

Perceptual-motor performance: messy writing but excellent artwork; poor eye–hand coordination, both gross and fine; mixes left–right direction, reverses digits, ambidextrous

Academic achievement: at grade level in mathematics but obtained grade equivalents of 11 and 12 for other subjects; rapid learner except in mathematics, excellent reading comprehension, uses odd syntax in written work—"To the store she went for to buy a gift for the little princess."

Social-emotional behavior: anxiety displayed during mathematics and language tests; sometimes tells of nightmares of bomb explosion when living in Israel

Parental involvement: mother teaches mathematics in same school; both parents show interest and are supportive of school program, show normal concern and expectations for Sali

Other comments: Sali appears to become more anxious in testing situations; areas that previously were her strong suits are being affected

SCHOOL PSYCHOLOGIST'S SUMMARY

Sali entered the testing room apprehensively and stood waiting to be asked to be seated. A fact-giving strategy established rapport within the first five minutes of the interview. She stated that she was on a diet because she wanted to win a bet with her girlfriend as to who could lose weight faster.

Her score on the WISC-R was in the superior range for the full IQ. The mean verbal score was significantly higher than the performance score. She showed strength in answering specific factual questions about informal educational settings. She understood verbal directions and showed awareness of cultural mores. She stated similarities in seemingly unrelated verbal stimuli and defined correctly over 90% of the vocabulary words on the test. Her recall of lists of digits, both forward and backward, was superior. In verbally solving arithmetic problems she did not score on the more complex items because of time; she answered correctly just after the time limit. She was able to identify missing parts in pictures and showed good comprehension of relationships when required to arrange pictures in sequence to tell a logical story. She missed even the simplest tasks involving manipulation of red and white blocks to represent a design to be copied. She could not put puzzles together to form completed objects, or associate two symbol systems and copy one on paper with pencil. She also had great difficulty in completing mazes, but when tested to limits without the time constraint she completed all mazes correctly. Her draw-a-man figure was elaborate and showed humor.

Sali's approach to tasks showed a deliberate, well-planned, problem-solving strategy. She used both hands on all tasks involving manipulation of objects and moved slowly and deliberately. She was well-informed and offered anecdotes about her experiences in Israel that related to vocabulary definitions and to information in general. She said that she often thought in Hebrew and then had to translate to English, thus slowing completion of tasks. She remarked that she sometimes translated to French or Russian for the fun of it, and this would slow up her classroom performance even more. She was cooperative and appeared pleased when allowed to complete tasks correctly after time limits were up. She handled failure on tasks with little frustration in situations where she was allowed to finish at her own speed. She began to breathe heavily and become flushed and excited when she was not allowed to complete a task beyond a time constraint.

During conversation at the end of the interview, she referred to how upset she was on the block design subtest of the WISC-R. She said that this happened on timed achievement tests and that she was glad they came only once a year. She stated that her parents and teachers would get all uptight when timed tests were to be given, and they got her so nervous she often would block—even before the test started. She stated that she appreciated their concern, as she knew that college entrance exams were timed, but she wanted eventually to do something related to art and did not believe her trouble with time would be a problem—if "they'd all get off my back."

In conclusion, Sali displayed mature social and emotional behavior. She showed insight into her strengths and weaknesses and enjoyed being in control of herself.

It appears that the significantly lower performance score may be due to time limits rather than severe perceptual disabilities. She did well on copying symbols but worked so slowly that her score was significantly lower in relationship to other manipulative tasks. The block design subtest was the first timed test, and she became uptight as soon as she noticed the stopwatch. She obviously was still anxious on the next test, arithmetic, which also was timed; this carried over to the third timed test in a row, assembling puzzles to form an object. Thus, it appears that characteristics inherent in the arrangement of the subtests may have provided data that usually are indicative of perceptual-motor disabilities. She did not reverse digits 0–9 when asked to write them.

It appeared that Sali was a gifted child but also highly creative, and these two characteristics do not always enhance performance on traditional school tasks. It was suggested that the figural form of the Torrance Test of Creative Thinking be administered to identify Sali's creative strengths and to aid in structuring her instructional program.

TEACHER'S SUMMARY OF INTERVIEW CHECKLIST (APPENDIX B)

Appearance: appropriate dress, facial expression is alert, attentive, happy; good physical health, a little too fat

Movement: observed—slow, clumsy, calm; inferred—purposeful, meaning, moves freely

Emotional: expression—pleasant, mostly relaxed, anxious during most timed tests, trusting; mood—steady, normal reaction; special symptoms—infrequent recurring nightmare of bomb exploding near apartment in Israel when she was 5

Orientation: spatial—knows where she is and how to get from one place to another; temporal—knows time of day, usually late

Verbal: talks a lot spontaneously, coherent conversation, stays on topic for appropriate time, unique use of tenses, high level of vocabulary, incorrect use of inflectional endings at times, unique syntax at times—especially in formal situations

Interpersonal relationships: peers—seeks out peers, is both leader and follower when appropriate, friends are her age and some adults, participates in group and sometimes withdraws from group to "do her own thing"; adults—friendly

Interactions in general: accepts rules reasonably; instructional program—joins in enthusiastically, shows initiative, suggests new ideas, handles frustration well, concentrates well, creative responses, interested, fluency of ideas in group discussions, flexible, original, tolerates ambiguity, resists premature closure

TEACHER'S SUMMARY OF IS-SLDM CHECKLIST (APPENDIX C)

Reasoning: strengths—sequences, errors made are logical, makes choices and decisions, connected cause–effect relationships, makes appropriate inferences from data and draws appropriate conclusions, no receptive language disabilities; weaknesses—reverses word meaning at times (right–left, in–out, top of–on bottom of), sometimes uses cues incorrectly—in computations goes from left to right; incorrect use of syntax but does comprehend meaning as in "John is lighter than Mary."

Problem solving: strengths—organizes tasks to facilitate completion, employs well-planned strategies; weakness—does not complete work on time and does not seem to care unless it bothers teacher or parents

Orientation: weaknesses—sometimes disoriented spatially, for example, laterality; often late returning from physical education class due to showering and changing clothes, computes +, ×, –algorithms in left–right direction

Motor performance: strengths—no excess tongue movement, salivation, or body movement under stress; good gross-motor and fine-motor coordination (especially in drawing minute details); weaknesses—messy writing, immature eye–hand coordination (cannot hit tennis ball even with eye glasses on), displays expressive language difficulties—seems to be sorting words before speaking in formal classroom situations

Attention: good

Perception: displays visual and auditory perception problems, such as reversing digits 3 and 7 at times, and seems to take a long time processing some words

Affect: displays inappropriate level of anxiety in relationship to task difficulty, especially in mathematics; often ambivalent regarding self-concept

TEACHER'S SUMMARY OF DEVELOPMENTAL MATHEMATICS CURRICULUM SCREENING CHECKLIST (APPENDIX D)

I. *Basic Relationships*: correctly performed all tasks, except writes mirror image of digits 3 and 7 at times

II. *Lower-Level Generalizations*: duality, set relationships, seriation, judging and estimating—correctly performed all tasks

III. *Concepts*: correctly performed all tasks

IV. *Higher-Level Relationships*: strengths—correctly completed all tasks up to mathematical language including inflectional endings, temporal, fractions; weaknesses—prepositions, missed conservation of time task

V. *Higher-Level Generalizations*: strengths—basic +, –, – facts, place-value notation, application to decimal notation, computation of whole numbers (+, ×, –) without renaming; multiplication, addition, subtraction of fractions with like and unlike denominators, addition and subtraction of decimals without renaming, addition and subtraction of mixed numbers without renaming, completed simple analogies, completed series of number patterns and of geometric figures, solved present-, past-, and future-tense word problems, cause–effect, topological equivalence, projective geometry, nonmetric geometry, metric geometry; can use graphs, can find mean, median, mode; uses and reads travel time schedules, can explain daylight saving time, meaning of *time and a half, double time*; explains and identifies different time zones in the world and can compute equivalent times across zones, uses time in cooking, correctly performed measurement tasks involving clothing size, temperature, metric measure; *weaknesses* — basic × facts; does not apply axioms, especially commutative and distributive properties; division of whole numbers with remaining in dividend; computation of whole numbers with renaming; division of fractions; multiplication of mixed numbers;

difficulty with mixed-tense and conditional-word problems; not exposed to network theory, probability; had not attempted to construct time schedules, late on school tasks

---

IEPM for Sali

Child's Name: Sali Pine   *Subject Area*: Mathematics
Level of Performance: performing at the top levels of the Developmental Mathematics Curriculum; achievement in mathematics, although at grade level, is much lower than her performance in other academic areas; highly creative

Prioritized Annual Goals:

I. To overcome excessive level of anxiety that blocks performance on timed mathematics tests
II. To learn to use mathematics language, especially inflectional endings and tense
III. To apply axioms in investigating higher-level generalizations
IV. To apply place value to computations involving renaming

Short-Term Objectives	IS	Evaluation	Persons Responsible	Length of Time
Annual Goal I				
Will construct time schedules for homework, duration of tasks	10.1.4	Keeps to schedule 80% of the time	Sali and parents	1 month
Will practice tasks in untimed situation first, and then perform same task in timed situation	9.5.1, 10.5.1, 12.1.3, 12.1.5	80% accuracy on times tasks	LD teacher and Sali	10 months
Annual Goal II				
Will correctly use language to express mathematical relationships	8.4.2– 8.4.8	90% accuracy	LD teacher	3 months
Annual Goal III				
Will identify axioms Will use axioms in computations	9.2.2, 9.1.7	” ”	Regular classroom teacher	3 months _ ”
Annual Goal IV				
Will apply place value to renaming	9.3.3, 9.3.4	90% accuracy	Regular classroom teacher	3 months

265

Summary: Sali

The members of the IEPM committee included Mr. Cohen, her fifth grade teacher; the LD teacher; Sali's parents; the school principal; the school psychologist; and Sali. The parents and Mr. Cohen agreed to "get off Sali's back" in timed test situations. The LD teacher and Sali agreed on the plan for helping her reduce tension on timed tests. Further observation by the LD teacher should focus on possible perceptual disability, since this phase of evaluation resulted in conflicting data.

The school personnel agreed that Sali performed best in open-ended situations, and ways were suggested for all concerned to use this approach with her. The Torrance method for teaching children with creative strengths (see Chapter 8, IS 8.3.6) was agreed upon as the major instructional approach.

Case Study: **Nancy**

Referral Form:

NAME Nancy Foster   AGE 12 years 3 months
SCHOOL P.S. 142   GRADE Special Class   SEX F
PARENTS/GUARDIAN Foster Mother: Juanita House
ADDRESS 4379 Molloy Road   PHONE 867-0932
REFERRED BY Ms. Harrington   POSITION County Social Worker
REASON FOR REFERRAL Foster mother's concern that Nancy is regressing; requests medical consultation

PROBLEM AREAS

Acculturation and language: does not speak
Health: thin, small, very small head
Hearing: appears OK
Vision: appears OK
Speech: makes sounds like an infant
Perceptual-motor performance: slow movements, drinks from cup, uses spoon and fork but otherwise displays no self-help skills; plays kickball and runs
Academic achievement: extremely slow rate of learning self-help skills
Social-emotional behavior: mostly appears to be in her own world, makes odd sounds
Parental involvement: Placed in institution for retarded (Developmental Center) at age 3 years 4 months. With foster mother for past three years (8 to 11 years of age). Foster mother is good to Nancy.
Other comments: Nancy will be enrolled in a special class in a public school next year. She has been attending the School for Retarded Citizens.

SUMMARY OF NEUROLOGICAL REPORT

Nancy was brought to the office by her foster mother with whom she had lived for three years. Her foster mother did not believe that Nancy has progressed significantly since coming to live with her family. She did not consistently understand verbal commands, although she followed what her foster mother said, such as "Go out in the other room" or "Open the door." For more elaborate commands, such as taking a cap off a pen, Nancy just kept moving the pen on the paper, even though no marks were made

266

because of the cap. She had minimal interaction, even with her foster mother. However, she did have some interaction when people looked straight in her eyes, and she smiled occasionally.

Nancy had four natural brothers. Her personal history indicated that she was a full-term baby and that delivery was normal. Nancy appeared to have developed normally until 4 months of age, when she developed a high fever that was accompanied with seizures. Teething did not begin at a normal age. She sat up at 2. At age 3, she said "Mama," "Daddy," and uttered sounds; crawled; and was able to stand and take a few steps while holding on to some support. She was not toilet trained at that age. Her seizures were controlled with medication.

She lived in a Developmental Center for children of ages 3 to 9. At the Center she was an ambulatory child who exhibited signs of hyperactivity. She did feed herself, and this was her only self-help skill. She had no speech but made sounds and was able to hum parts of songs she apparently had memorized. She was now toilet trained, although she did continue to have some accidents.

Nancy did not make any articulated sounds, although she did make repetitive noises with her mouth. She liked to play kickball. She did not watch television.

On examination, she was a thin and small girl with a head circumference of 48 cm, which was well below two standard deviations. Her neurological examination showed otherwise normal development of eye pupils, extremities, reflexes, and sensations.

In conclusion, Nancy showed severe developmental delay in several areas. In speech she functioned below 1 year of age. In social adaptive skills she functioned around 1 to 1½ years. In motor areas she functioned between 2 and 3 years. She also had a history of seizures that she had outgrown and for which she no longer took medication. In addition to her severe developmental delay and lack of speech, many of her behavior characteristics had an autistic quality. This made it difficult to approach her and to establish any continuous communication with her.

SUMMARY OF INTERVIEW CHECKLIST BY FAMILY PHYSICIAN

Appearance: appropriate dress; facial expression—placid, attentive, happy; physical health—fair, too thin

Movement: observed—slow, clumsy, restless, mannerisms with arms and hands; inferred—no purpose, no meaning, inhibited

Emotional: expression—pleasant, anxious, suspicious; mood—fluctuates, overreaction; special symptoms—misinterprets environmental cues

Orientation: spatial—knows where she is, confused as to how to get from one place to another; temporal—confused as to time of day, usually late

Verbal: does not talk

Interpersonal relationships: peers—isolate, follower, withdraws from groups; adults—fearful

Interactions in general: rules—shows passive resistance, behavior at times openly rebellious; instructional program—disruptive, needs direction, handles frustrations poorly, distractible, preoccupied, rigid, needs structure

SOCIAL WORKER/FOSTER MOTHER SUMMARY OF DEVELOPMENTAL MATHEMATICS CURRICULUM CHECKLIST (APPENDIX D)

I. *Basic Relationships: strengths*—succession—can follow simple directions, e.g., "open the door, go into the next room," walks in normal manner, performs some self-help skills in proper succession, e.g., eating, toileting; topological relationship—can stand close to foster mother on command from her; enclosure—can place object inside a box (puts lollipops in her cloth bag that she carries around with her)

II. *Lower-Level Generalizations: strengths*—sorting—attempted to use a pen given her by neurologist during his examination as she uses a crayon, having had no previous experience with a pen with a cap on it

III. *Concepts: strengths*—shown a penny and a truck, can pick up the penny on command

---

IEPM for Nancy

Child's Name: Nancy Foster   *Subject Area*: Mathematics

Level of Performance: functioning at 1–3-year level in all areas; appears removed from reality at times; knows where she is spatially but confused as to how to get from one place to another; no verbal communication; can perform some basic relationship tasks such as succession, proximity, and enclosure; makes simple lower-level generalizations; can select a coin where a simple choice must be made

Prioritized Long-Term Goals:

I. To use alternative symbol system for communication
II. To use basic mathematics relationships in simple household tasks in the classroom and home
III. To use money in real-life situations
IV. To use lower-level generalizations in simple tasks

IEP Goal: To improve interpersonal relationships

Short-Term Objectives	IS	Special Educational And/or Support Services	Persons Responsible	Length of Time
Annual Goal I				
To use pictographs	8.4.1– 8.4.5	Language disorders specialist in computerized symbol communication	EMR teacher	10 months
Annual Goal II				
To pay fare to ride bus to school	12.3.3		Foster grandmother	1 month
To pay 1 penny (given by teacher) for 1 crayon in classroom	5.1.1		EMR teacher	1 week
Annual Goal III				
To set table for juice giving 1 cup, 1 napkin to each of 3 others	5.1.1		EMR teacher	1 month

Short-Term Objectives	IS	Special Educational and/or Support Services	Persons Responsible	Length of Time
To arrange boots neatly in succession on floor	5.1.1		"	1 month
To stack plastic saucers on top of plastic dinner plates	7.1.1		"	1 month
Annual Goal IV				
To pay 2 pennies for 1 tootsie roll	5.1.3, 6.5.1, 9.1.6		"	1 week
To pay 5 pennies for 1 cup of soup	6.5.2		"	1 month
To exchange 5 pennies for 1 nickel and use either form to pay for 1 cup of soup, milk, pop, etc.	6.5.2, 8.3.1, 8.3.5		"	6 months
IEP Annual Goal				
To interact with others	12.1.5, 12.2.1, 12.3.2	Social work services	Older persons working with special children program	12 months

Summary: Nancy

The IEPM committee comprised Ms. Harrington, the social worker who made the referral on Nancy; Mrs. House, the foster mother; the principal of the public school where she will be placed in the fall; the EMR teacher in her new school, and the school psychologist for the EMR-BD program in the school, who represented the pediatric neurologist and Nancy's family physician. The checklists showed that Nancy had many strengths related to basic components of the Developmental Mathematics Curriculum. The IEPM committee agreed to use these strengths in a systematic program.

Since the foster mother was concerned that Nancy had not progressed since she came to live with her family, it was agreed that periodic committee evaluations would occur with frequency—at one-month intervals until some progress was noted.

A priority goal was to help Nancy improve her interpersonal relationships. The community Foster Grandparent Program was contacted, and one of their volunteers agreed to ride the bus with Nancy, thus helping with both the broad IEP goal and Annual Goal III.

269

G Cross-Reference Outline By Instructional Strategies

Instructional Strategy	Related Generic Influence on Learning	Mathematical Topic Particularly Appropriate
5.1.1 Present small amounts; chunking	1.1.0 Rate and amount 1.7.0 Size of vocabulary 1.9.0 Attend to salient features 1.13.0 Ability to cope with complex 2.1.0 Disorders of visual perception 2.2.0 Disorders of auditory perception	5.1.0 Sequencing 5.2.0 One-to-one correspondence 6.2.0 Equivalence of sets 6.4.0 Sorting 6.6.0 Seriation 6.7.0 Judging
5.1.2 Incorporate redundancy	1.1.0 Rate and amount 1.2.0 Speed of learning related to specific content 1.3.0 Ability to retain information 1.4.0 Need for repetition 1.6.0 Ability to learn arbitrary symbols 2.1.4 Deficit in visual-sequential memory 2.2.5 Problems in auditory sequential memory	5.1.0 Sequencing 5.4.0 Arbitrary associations 6.1.0 Duality 6.4.0 Sorting
5.1.3 Use of cues	1.6.0 Ability to learn arbitrary associations 1.9.0 Ability to attend to salient aspects of a situation 2.1.0 Disorders of visual perception 2.1.1 Poor visual discrimination 2.1.2 Visual figure-ground distractibility 2.1.3 Form constancy problems 2.1.5 Difficulties in spatial relationships 2.2.0 Disorders of auditory perception	

Instructional Strategy	Related Generic Influence on Learning		Mathematical Topic Particularly Appropriate	
5.1.3 Use of cues contd.	2.2.1	Poor auditory discrimination	5.1.0	Sequencing
	2.2.2	Auditory figure-ground distractibility	5.3.0	Topological relationships
	4.3.0	Immaturity	6.4.0	Sorting
	4.4.0	Deficiencies in moral development	6.5.0	Many-to-one and one-to-many matching
			6.6.0	Seriation
			7.1.0	Concepts
			8.1.0	Transformation
			9.7.0	Cause–effect
5.1.4 Separating and underlining	1.6.0	Ability to learn symbols	5.1.0	Sequencing
	1.9.0	Ability to attend to salient aspects	5.3.0	Topological relationships
	2.1.0	Disorders of visual perception	5.4.0	Arbitrary associations
	2.1.1	Poor visual discrimination		
	2.1.2	Visual figure–ground distractibility		
	2.1.5	Difficulties in spatial relationships		
5.1.5 Control number of dimensions that define a linear sequence	1.1.0	Rate and amount of learning	5.1.0	Sequencing
	1.2.0	Speed of learning related to specific content	5.2.0	One-to-one correspondence
	1.3.0	Ability to retain information	6.1.0	Duality
	1.9.0	Ability to attend to salient aspects	6.2.0	Equivalence of sets
	1.11.0	Ability to make decisions and judgments	6.3.0	Greater than–less than matching
	2.1.0	Disorders of visual perception	6.4.0	Sorting
	2.1.1	Poor visual discrimination	6.6.0	Seriation
	2.1.2	Visual figure–ground distractibility	6.7.0	Judging and estimating
	2.1.3	Form constancy problems	8.1.0	Transformations
	2.2.1	Poor auditory discrimination	8.2.0	Set operations and relationships
	2.2.2	Auditory figure–ground distractibility	8.3.0	Number operations and relationships

	Instructional Strategy		Related Generic Influence on Learning		Mathematical Topic Particularly Appropriate
5.1.6	Emphasize patterns	1.3.0	Ability to retain information	5.1.0	Sequencing
		1.4.0	Need for repetition	5.2.0	One-to-one corre-spondence
		1.5.0	Verbal skills	5.3.0	Topological rela-tionships
		1.6.0	Ability to learn arbi-trary associations	5.4.0	Arbitrary asso-ciations
		1.8.0	Ability to form relationships	6.2.0	Equivalence of sets
		1.9.0	Ability to attend to salient aspects	6.6.0	Seriation
		1.10.0	Use of problem-solving strategies	7.1.0	Concepts
		2.1.0	Disorders of visual perception	8.1.0	Transformations
		2.1.1	Poor visual discrimi-nation	8.2.0	Set relationships
		2.1.2	Visual figure–ground distractibility	8.3.0	Number relationships
		2.1.3	Form constancy problems	9.1.0	Basic facts
		2.1.4	Deficit in visual-sequential memory	9.2.0	Axioms
		2.1.5	Difficulties in spatial relationships	9.3.0	Place value
		2.2.0	Disorders of auditory perception	9.5.0	Seriation extended
		2.2.1	Poor auditory dis-crimination	9.12.0	Network theory
		2.2.2	Auditory figure-ground distractibility	9.14.0	Applied mathe-matics—measurement
5.1.7	Introduce additional relevant dimensions for generating an alternating pattern	1.8.0	Ability to form relationships and generalizations	5.1.0	Sequencing
		1.9.0	Ability to attend to salient aspects	5.2.0	One-to-one correspondence
		1.12.0	Ability to draw infer-ences and hypothesize	6.6.0	Seriation
		1.13.0	Ability in general to abstract	7.1.0	Concepts
				9.5.0	Seriation extended
				9.14.0	Applied mathematics
5.1.8	Use rehearsal strategies	1.3.0	Ability to retain information	5.1.0	Sequencing
		1.4.0	Need for repetition	5.4.0	Arbitrary associations
		1.5.0	Verbal skills	6.7.0	Judging and estimating
				9.1.0	Basic facts

Instructional Strategy	Related Generic Influence on Learning	Mathematical Topic Particularly Appropriate
5.1.8 Use rehearsal strategies contd.	1.6.0 Ability to learn arbitrary associations and symbol systems 1.10.0 Use of problem-solving strategies 2.1.4 Deficit in visual-sequential memory 2.2.5 Problems in auditory sequential memory	
5.1.9 Provide complex and/or subtle sequencing activities	1.1.0 Rapid rate of learning 1.3.0 Ability to retain information 1.8.0 Ability to form relationships, concepts, and generalizations 1.9.0 Ability to attend to salient aspects of a situation 1.11.0 Ability to make decisions and judgments 1.12.0 Ability to draw inferences and conclusions 1.13.0 Ability to abstract and cope with complexity	5.1.0 Sequencing 6.6.0 Seriation 6.7.0 Judging and estimating 9.5.0 Seriation extended 9.7.0 Cause–effect extended
5.1.10 Incorporate incompleteness to activate creative potential	1.2.0 Speed of learning related to specific content 1.8.0 Ability to form relationships, concepts, and generalizations 1.9.0 Ability to attend to salient aspects of a situation 1.10.0 Use of problem-solving strategies 1.12.0 Ability to draw inferences and conclusions, and to hypothesize 1.13.0 Ability in general to abstract	5.1.0 Sequencing 6.7.0 Judging and estimating 8.1.0 Transformations 9.5.0 Seriation extended 9.6.0 Solving word problems 9.7.0 Cause–effect extended 9.8.0 Transformations-topological 9.11.0 Geometry (metric) 9.12.0 Network theory 9.13.0 Probability, statistics, graphs
5.1.11 Producing something and using it	1.7.0 Allows for expansion of size of vocabulary since child describes personal product 1.10.0 Use of problem solving strategies	5.1.0 Sequencing 8.1.0 Transformations

Instructional Strategy	Related Generic Influence on Learning	Mathematical Topic Particularly Appropriate
5.2.1 Employ frequency to develop number word-object associations	1.1.0 Rate of learning 1.3.0 Ability to retain information 1.4.0 Need for repetition 1.5.0 Verbal skills 1.6.0 Ability to learn arbitrary associations and symbol systems 1.7.0 Size of vocabulary 2.2.0 Disorders of auditory perception 2.3.0 Rules of general language and mathematics	5.2.0 One-to-one correspondence 5.4.0 Arbitrary associations 8.4.0 Mathematics as a language of relationships 9.6.0 Solving word problems 9.14.0 Applied mathematics—measurement
5.2.2 Employ familiarity to develop number word-object associations	Same as IS 5.2.1	Same as IS 5.2.1
5.2.3 Employ vividness or imagery to develop number word-object association	1.3.0 Ability to retain information 1.4.0 Need for repetition 1.6.0 Ability to learn arbitrary associations 1.9.0 Ability to attend to salient aspects of a situation 2.1.4 Deficits in visual memory 2.2.5 Problems in auditory memory 2.3.0 Rules of general language and mathematics	5.2.0 One-to-one correspondence 5.4.0 Arbitrary associations 8.4.0 Mathematics as a language of relationships 9.6.0 Solving word problems 9.14.0 Applied mathematics
5.2.4 Emphasize patterns that generate sequences to be matched	1.1.0 Rate of learning 1.6.0 Ability to learn arbitrary associations 1.8.0 Ability to form relationships, concepts, and generalizations 1.9.0 Ability to attend to salient aspects of a situation 1.11.0 Ability to make decisions and judgments 2.1.0 Disorders of visual perception 2.1.1 Poor visual discrimination	5.1.0 Sequencing 5.2.0 One-to-one correspondence 5.4.0 Arbitrary associations 6.2.0 Equivalence of sets 6.3.0 Greater than-less than matching 6.5.0 Many-to-one and one-to-many 6.6.0 Seriation 6.7.0 Judging and estimating

Instructional Strategy	Related Generic Influence on Learning	Mathematical Topic Particularly Appropriate
5.2.4 Emphasize patterns that generate sequences to be matched contd.	2.1.2 Visual figure–ground distractibility 2.1.3 Form constancy problems 2.1.4 Deficit in visual-sequential memory 2.1.5 Difficulties in spatial relationships	
5.3.1 Reinforce attention to a relevant dimension	1.8.0 Ability to form relationships, concepts, and generalizations 1.9.0 Ability to attend to salient aspects of a situation 2.1.1 Poor visual discrimination 2.1.2 Visual figure–ground distractibility 2.1.3 Form constancy problems 2.1.4 Deficit in visual-sequential memory 2.1.5 Difficulties in spatial relationships 2.2.1 Poor auditory discrimination 2.2.2 Auditory figure–ground distractibility 2.2.5 Problems in auditory sequential memory	5.1.0 Sequencing 5.2.0 One-to-one correspondence 5.3.0 Topological relationships 5.4.0 Arbitrary shapes of symbols 6.2.0 Equivalence of sets 6.4.0 Sorting 6.6.0 Seriation 6.7.0 Judging 7.1.0 Concepts 8.1.0 Transformations 8.2.0 Set relationships 8.4.0 Mathematics as a language of relationships 9.8.0 Transformations-topological 9.10.0 Geometry (nonmetric, metric) 9.12.0 Network theory 9.14.0 Applied mathematics
5.3.2 Point out relevant relationships	1.1.0 Rate and amount of learning 1.8.0 Ability to form relationships, concepts, and generalizations 1.9.0 Ability to attend to salient aspects of a situation 1.13.0 Ability in general to abstract and cope with complexity 2.1.0 Disorders of visual perception 2.2.0 Disorders of auditory perception 3.4.0 Sensory limitation	5.0.0 Pre-number relationships 5.3.0 Topological relationships 6.2.0 Equivalence of sets 6.3.0 Greater than-less than matching 6.4.0 Sorting 6.6.0 Seriation 8.2.0 Set relationships 8.3.0 Number relationships 8.4.0 Mathematics as a language of relationships

Instructional Strategy	Related Generic Influence on Learning		Mathematical Topic Particularly Appropriate	
5.3.2 Point out relevant relationships contd.			9.2.0	Axioms
			9.3.0	Place value
			9.6.0	Solving word problems (simple syntax)
			9.7.0	Cause–effect extended
			9.14.0	Applied mathematics
			9.15.0	Economics
5.3.3 Incorporate movement	1.1.0	Rate and amount of learning	5.2.0	One-to-one correspondence
	1.8.0	Ability to see relationships	5.3.0	Topological relationships
	2.0.0	Psychomotor abilities or motor effects of mental processes		
	3.1.0	Physical impairment		
	3.2.0	Low-vitality and fatigue conditions		
	3.3.0	Sensory limitation		
5.4.1 Emphasize differences in distinctive features of stimuli		Same as ISs 5.1.3, 5.1.5, 5.1.7, 5.3.1	5.3.0	Topological relationships
			5.4.0	Arbitrary associations
			8.2.0	Set relationships
			8.4.0	Mathematics as a language of relationships
			9.1.0	Basic facts
			9.8.0	Transformations-topological
			9.10.0	Geometry (nonmetric)
			9.14.0	Applied mathematics—measurement
5.4.2 Control irrelevant stimuli		Same as IS 5.1.5	5.1.0	Sequencing
	1.6.0	Ability to learn arbitrary associations and symbol systems	5.4.0	Arbitrary associations
			7.1.0	Concepts

Instructional Strategy	Related Generic Influence on Learning	Mathematical Topic Particularly Appropriate
5.4.3 Use the student's preferred dimension	1.1.0 Rate of learning (slow) 1.3.0 Ability to retain information 1.6.0 Ability to learn arbitrary associations 2.1.1 Poor visual discrimination 2.1.2 Visual figure–ground distractibility	5.4.0 Arbitrary associations
5.4.4 Place stimuli to be discriminated in close proximity on initial tasks	1.1.0 Rate of learning 1.6.0 Ability to learn arbitrary associations	5.4.0 Arbitrary associations
5.4.5 Increase the number of relevant dimensions	Same as IS 5.1.3 1.1.0 Rate of learning (slow)	5.4.0 Arbitrary associations
5.4.6 Use cues that are different during initial learning	Same as IS 5.1.3	5.4.0 Arbitrary associations
5.4.7 Include identical stimuli to serve as background and thus to aid attention to relevant cues	1.1.0 Rate of learning (slow) 1.6.0 Ability to learn arbitrary associations 1.9.0 Ability to attend to salient aspects of a situation 2.1.0 Disorders of visual perception 2.1.2 Visual figure–ground distractibility	5.4.0 Arbitrary associations
5.4.8 Reinforce alternative positions on tasks involving random response patterns	1.1.0 Rate of learning (slow) 1.9.0 Ability to attend to salient aspects of a situation	5.4.0 Arbitrary associations

Instructional Strategy	Related Generic Influence on Learning		Mathematical Topic Particularly Appropriate	
6.1.1 Employ familiarity to develop duality	1.8.0	Ability to form relationships, concepts, and generalizations	6.1.0	Duality
	1.10.0	Use of problem-solving strategies		
6.1.2 Point out relevant relationship	1.9.0	Ability to attend to salient aspects of a situation	5.1.0 5.3.0	Sequencing Topological relationships
	1.12.0	Ability to draw inferences and conclusions	6.1.0 6.2.0 6.3.0 6.4.0 8.4.0	Duality Equivalence of sets Greater-less than Sorting Mathematics as a language of relationships
6.1.3 Use redundancy and point out relevant dimension	1.9.0	Ability to attend to salient aspects of a situation		Same as IS 6.1.2
6.2.1 Control the number of dimensions embedded in the stimuli	1.8.0	Ability to form relationships, concepts, and generalizations	6.2.0 7.1.0 9.4.0	Equivalence of sets Concepts Computation
	1.9.0	Ability to attend to salient aspects of a situation		
	2.1.0	Disorders of visual perception		
6.2.2 Use chunking to control distractions in the ground	2.1.2	Visual figure–ground distractibility	6.2.0 6.6.0 9.4.0	Equivalence of sets Seriation Computation
	2.2.2	Auditory figure-ground distractibility		
6.2.3 Prevent incorrect responses	1.11.0	Ability to make decisions and judgments	6.2.0 6.7.0	Equivalence of sets Judging and estimating
	4.1.0	Hyperactivity, distractibility, impulsivity		
6.3.1 Restrict complexity of incidental learning to the gross and obvious	1.9.0	Ability to attend to salient aspects of a situation	6.1.0 6.2.0 6.3.0 6.4.0 6.6.0 7.1.0	Duality Equivalence of sets Greater-less than Sorting Seriation Concepts
	2.1.1	Poor visual discrimination		
	2.2.1	Poor auditory discrimination		

278

Instructional Strategy	Related Generic Influence on Learning	Mathematical Topic Particularly Appropriate
6.3.2 Encourage consistent responses to similar stimuli	1.3.0 Ability to retain information 1.4.0 Need for repetition 2.1.0 Disorders of visual perception 2.2.0 Disorders of auditory perception 4.4.0 Deficiencies in moral development	6.3.0 Greater-less than 6.4.0 Sorting
6.3.3 Replace incidental learning tasks by structured intentional learning tasks	1.8.0 Ability to form relationships, concepts, and generalizations 1.9.0 Ability to attend to salient aspects of a situation 1.10.0 Use of problem-solving strategies 1.11.0 Ability to make decisions and judgments 1.13.0 Ability in general to abstract and cope with complexity 2.1.3 Form-constancy problems 4.1.0 Hyperactivity, distractibility, impulsivity 4.4.0 Deficiencies in moral development	6.3.0 Greater-less than 6.4.0 Sorting 6.7.0 Judging and estimating
6.4.1 Reduce complexity of task	1.1.0 Rate and amount 1.8.0 Ability to form relationships, concepts, and generalizations 1.9.0 Ability to attend to salient aspects of a situation 2.1.0 Disorders of visual perception 4.1.0 Distractibility	6.4.0 Sorting 6.7.0 Judging and estimating
6.4.2 Use the familiar and useful	1.13.0 Ability in general to abstract and cope with complexity 2.1.3 Form-constancy problems 4.3.0 Inadequacy	6.4.0 Sorting

279

Instructional Strategy	Related Generic Influence on Learning		Mathematical Topic Particularly Appropriate	
6.4.3 Use the strategy of becoming more familiar with the familiar	1.10.0 3.3.0	Use of problem-solving strategies Sensory limitation	6.4.0 7.1.0	Sorting Concepts
6.5.1 Use consistent vocabulary	1.1.0 1.2.0 1.3.0 1.4.0 1.5.0 1.6.0 1.7.0 3.3.0 4.1.0	Rate and amount Speed of learning/ content Ability to retain information Need for repetition Verbal skills Ability to learn arbitrary associations and symbol systems Size of vocabulary Sensory limitation Hyperactivity, distractibility, impulsivity	6.3.0 6.5.0 8.4.0 9.3.0 9.6.0	Greater-less than Many-to-one and one-to-many matching Mathematics as a language of relationships Place value Solving word problems
6.5.2 Restrict complexity of the task and incorporate preference into stimuli	1.13.0 3.3.0 4.3.0	Ability in general to abstract and cope with complexity Sensory limitation Withdrawal	6.4.0 6.5.0 7.1.0	Sorting Many-to-one and one-to-many matching Concepts
6.6.1 Use a model whose competence in the task has been established	4.1.0 4.2.0 4.3.0 4.4.0	Impulsivity Aggression Withdrawal, immaturity, inadequacy Deficiencies in moral development	6.6.0 6.7.0 9.14.0 9.15.0	Seriation Judging and estimating Applied mathematics—measurement Applied mathematics—economics
6.6.2 Use of verbalization by model	1.5.0 3.3.0 4.2.0	Verbal skills Sensory limitation Aggression	6.6.0 6.7.0 8.4.0 9.15.0	Seriation Judging and estimating Mathematics as a language of relationships Applied mathematics—economics

280

Instructional Strategy	Related Generic Influence on Learning		Mathematical Topic Particularly Appropriate	
6.6.3 Control the number of dimensions embedded in the stimuli	1.9.0	Ability to attend to salient aspects of a situation	6.4.0	Sorting
	2.1.0	Disorders of visual perception	6.6.0	Seriation
	2.2.0	Disorders of auditory perception	7.1.0	Concepts
	3.3.0	Sensory limitation		
	4.1.0	Distractibility		
6.7.1 Restrict complexity of the stimuli	1.9.0	Ability to attend to salient aspects of a situation	6.7.0	Judging and estimating
	1.13.0	Ability in general to abstract and cope with complexity	7.1.0	Concepts
			8.2.0	Set operations and relationships
			8.4.0	Mathematics as a language of relationships
	3.3.0	Sensory limitation	9.4.0	Computation
	4.1.0	Distractibility	9.6.0	Solving word problems
6.7.2 Provide sequencing experiences comprised of more abstract and/or subtle nature	1.8.0	Ability to form relationships, concepts, and generalizations	6.6.0	Seriation
			6.7.0	Judging and estimating
	1.13.0	Ability in general to abstract and cope with complexity	9.5.0	Seriation extended
6.7.3 Encourage deferred judgment during problem-solving activity	1.3.0	Ability to retain information	6.7.0	Judging and estimating
	1.10.0	Use of problem-solving strategies	9.7.0	Cause–effect extended
	1.11.0	Ability to make decisions and judgments	9.15.0	Applied mathematics—economics
	4.1.0	Impulsivity		
	4.4.0	Deficiencies in moral development		
7.1.1 Use familiar categories to insure meaning for the student	1.4.0	Need for repetition	7.1.0	Concepts
	3.3.0	Sensory limitation	9.6.0	Solving word problems
	4.3.0	Withdrawal, immaturity, inadequacy		
7.1.2 Provide bimodal input	3.3.0	Sensory limitation	6.7.0	Judging and estimating
			7.1.0	Concepts

Instructional Strategy	Related Generic Influence on Learning		Mathematical Topic Particularly Appropriate	
7.1.3 Group input by category	1.8.0	Ability to form relationships, concepts, and generalizations	6.4.0	Sorting
	1.9.0	Ability to attend to salient aspects of a situation	7.1.0	Concepts
	4.1.0	Distractibility		
7.1.4 Increase task complexity	1.8.0	Ability to form relationships, concepts, and generalizations	6.6.0	Seriation
			6.7.0	Judging and estimating
	1.9.0	Ability to attend to salient aspects of a situation	7.1.0	Concepts
			9.5.0	Seriation extended
			9.6.0	Solving word problems
			9.13.0	Probability, statistics, graphs
7.1.5 Employ creative problem-solving strategy to facilitate concept formation	1.8.0	Ability to form relationships, concepts, and generalizations	7.1.0	Concepts
			9.15.0	Applied mathematics—economics
	1.10.0	Use of problem-solving strategies		
	1.12.0	Ability to draw inferences and conclusions		
8.1.1 Use of modified conservation tasks	1.9.0	Ability to attend to salient aspects of a situation	8.1.0	Transformations
			9.8.0	Topological equivalence
	1.10.0	Use of problem-solving strategies	9.14.0	Applied mathematics—measurement
	1.12.0	Ability to draw inferences and conclusions		
	3.3.0	Sensory limitation		
	4.1.0	Distractibility		
8.1.2 Use of peer team learning	1.4.0	Need for repetition		All topics
	1.5.0	Verbal skills		
	1.6.0	Ability to learn arbitrary associations and symbol systems		
	1.7.0	Size of vocabulary		
	1.9.0	Ability to attend to salient aspects of a situation		
	3.1.0	Physical impairments		
	3.2.0	Low vitality and fatigue conditions		
	3.3.0	Sensory limitation		

Instructional Strategy	Related Generic Influence on Learning	Mathematical Topic Particularly Appropriate
8.1.3 Apply modified conservation tasks to rigid transformations	Same as IS 8.1.1	8.1.0 Transformations 9.8.0 Topological equivalence
8.1.4 Employ creative problem solving	1.8.0 Ability to form relationships, concepts, and generalizations 1.10.0 Use of problem-solving strategies 3.3.0 Sensory limitation	6.7.0 Judging and estimating 8.1.0 Transformation 9.4.0 Computation 9.8.0 Topological equivalence 9.9.0 Projective geometry 9.12.0 Network theory 9.13.0 Probability, statistics, graphs
8.2.1 Control for component skills	1.1.0 Rate and amount 1.2.0 Speed of learning/content 1.5.0 Verbal skills 1.9.0 Ability to attend to salient aspects of a situation	All topics
8.2.2 Use natural categories for classification tasks rather than arbitrary categories	1.8.0 Ability to form relationships, concepts, and generalizations 1.9.0 Ability to attend to salient aspects of a situation 1.11.0 Ability to make decisions and judgments 1.13.0 Ability in general to abstract and cope with complexity	7.1.0 Concepts 8.2.0 Set operations and relationships
8.2.3 Distinguish between intensional and extensional aspects of concepts	Same as IS 8.2.2	Same as IS 8.2.2
8.2.4 Distinguish between classes and collections	1.11.0 Ability to make decisions and judgments 1.13.0 Ability in general to abstract and cope with complexity	7.1.0 Concepts 8.2.0 Set operations and relationships

283

Instructional Strategy	Related Generic Influence on Learning	Mathematical Topic Particularly Appropriate
8.2.5 Use good exemplars of a class	1.13.0 Ability in general to abstract and cope with complexity 3.3.0 Sensory limitation	Same as IS 8.2.4
8.2.6 Restrict the number of exemplars in classification tasks	1.13.0 Ability in general to abstract and cope with complexity 4.1.0 Distractibility	Same as IS 8.2.4
8.3.1 Provide immediate knowledge of results	4.3.0 Withdrawal, immaturity, inadequacy	All topics
8.3.2 Use distributed repetition	1.1.0 Rate and amount 1.2.0 Speed of learning/content 1.3.0 Ability to retain information 1.4.0 Need for repetition 1.6.0 Ability to learn arbitrary associations and symbol systems 3.2.0 Low-vitality and fatigue conditions 4.1.0 Hyperactivity	All topics
8.3.3 Use of verbal rehearsal	1.4.0 Need for repetition 1.6.0 Ability to learn arbitrary associations and symbol systems 1.9.0 Ability to attend to salient aspects of a situation 2.1.0 Disorders of visual perception 2.1.4 Deficit in visual-sequential memory 2.2.5 Problem in auditory sequential memory	6.4.0 Sorting 6.6.0 Seriation 8.3.0 Number operations and relationships
8.3.4 Facilitate attention to the relevant dimension	1.9.0 Ability to attend to salient aspects of a situation 2.1.1 Poor visual discrimination 2.2.1 Poor auditory discrimination 4.1.0 Distractibility	Same as IS 6.1.2

Instructional Strategy	Related Generic Influence on Learning		Mathematical Topic Particularly Appropriate	
8.3.5 Plan for transfer in learning	1.8.0	Ability to form relationships, concepts, and generalizations	All topics	
	1.12.0	Ability to draw inferences and conclusions		
8.3.6 Employ Torrance's Hierarchy of Creative Skills	1.10.0	Use of problem-solving strategies	6.7.0	Judging and estimating
	4.4.0	Deficiencies in moral development	7.1.0	Concepts
			8.1.0	Transformations
			8.2.0	Set operations and relationships
			8.3.0	Number operations and relationships
			8.4.0	Mathematics as a language of relationships
			9.5.0	Seriation extended
			9.15.0	Applied mathematics—economics
8.4.1 Use shorter, simpler sentences when giving directions	1.1.0	Rate and amount	All topics	
	1.2.0	Speed of learning/content		
	1.3.0	Ability to retain information		
	1.4.0	Need for repetition		
	1.5.0	Verbal skills		
	1.7.0	Size of vocabulary		
	1.9.0	Ability to attend to salient aspects of a situation		
	1.13.0	Ability in general to abstract and cope with complexity		
	2.2.1	Poor auditory discrimination		
	4.1.0	Hyperactivity, distractibility, impulsivity		

Instructional Strategy	Related Generic Influence on Learning	Mathematical Topic Particularly Appropriate
8.4.2 Use concrete examples of spatial and quantitative relationships	1.8.0 Ability to form relationships, concepts, and generalizations 1.9.0 Ability to attend to salient aspects of a situation 1.13.0 Ability in general to abstract and cope with complexity 3.3.0 Sensory limitation 4.4.0 Deficiencies in moral development	6.3.0 Greater-less than 6.5.0 Many-to-one and one-to-many matching 6.7.0 Judging and estimating 7.1.0 Concepts 8.1.0 Transformations 8.2.0 Set operations and relationships 8.3.0 Number operations and relationships 8.4.0 Mathematics as a language of relationships 9.1.0 Basic facts 9.2.0 Axioms 9.3.0 Place value 9.4.0 Computation 9.8.0 Topological equivalence
8.4.3 Use prompting	1.1.0 Rate and amount 1.8.0 Ability to form relationships, concepts, and generalizations 1.9.0 Ability to attend to salient aspects of a situation 2.1.4 Deficit in visual-sequential memory	All topics
8.4.4 Use relational rebus symbols	1.6.0 Ability to learn arbitrary associations and symbol systems 1.7.0 Size of vocabulary 1.8.0 Ability to form relationships, concepts, and generalizations 3.3.0 Sensory limitation	8.4.0 Mathematics as a language of relationships

Instructional Strategy	Related Generic Influence on Learning	Mathematical Topic Particularly Appropriate
8.4.5 Incorporate modeling plus verbal instructions	1.5.0 Verbal skills 2.3.1 Rules of general language 4.1.0 Hyperactivity, distractibility, impulsivity 4.2.0 Aggression 4.3.0 Withdrawal, immaturity, inadequacy 4.4.0 Deficiencies in moral development	5.4.0 Arbitrary association 8.4.0 Mathematics as a language of relationships 9.6.0 Solving word problems
8.4.6 Highlight inflectional endings	1.6.0 Ability to learn arbitrary associations and symbol systems 1.9.0 Ability to attend to salient aspects of a situation	5.4.0 Arbitrary association 8.4.0 Mathematics as a language of relationships
8.4.7 Introduce novelty into learning experience	1.10.0 Use of problem-solving strategies	All topics
8.4.8 Show how different verbal directions may change spatial relationships	1.5.0 Verbal skills 1.8.0 Ability to form relationships, concepts, and generalizations	8.4.0 Mathematics as a language of relationships
9.1.1 Emphasize patterns	1.4.0 Need for repetition 1.9.0 Ability to attend to salient aspects of a situation 2.1.0 Disorders of visual perception 2.2.0 Disorders of auditory perception 3.3.0 Sensory limitation 4.1.0 Hyperactivity, distractibility, impulsivity	5.1.0 Sequencing 6.6.0 Seriation 9.1.0 Basic facts
9.1.2 Use rehearsal strategies	1.3.0 Ability to retain information 1.4.0 Need for repetition 2.2.5 Problems in auditory sequential memory 4.1.0 Hyperactivity, distractibility, impulsivity	9.1.0 Basic facts

Instructional Strategy	Related Generic Influence on Learning	Mathematical Topic Particularly Appropriate
9.1.3 Emphasize differences in distinctive features of stimuli	1.9.0 Ability to attend to salient aspects of a situation 2.1.1 Poor visual discrimination 2.2.1 Poor auditory discrimination 3.3.0 Sensory limitation	5.1.0 Sequencing 6.2.0 Equivalence of sets 6.4.0 Sorting 6.6.0 Seriation 7.1.0 Concepts 8.1.0 Transformations 9.1.0 Basic facts 9.2.0 Axioms 9.3.0 Place value
9.1.4 Restrict size of sum	1.1.0 Rate and amount 1.2.0 Speed of learning/ content	9.1.0 Basic facts 9.4.0 Computation
9.1.5 Use R-K counting boards	1.2.0 Speed of learning/ content 1.8.0 Ability to form relationships, concepts, and generalizations 1.10.0 Use of problem-solving strategies	8.3.0 Number operations and relationships 9.1.0 Basic facts 9.3.0 Place value 9.4.0 Computation
9.1.6 Make 2-for-1 exchanges on R-K counting board	Same as IS 9.1.5	Same as IS 9.1.5
9.1.7 Use Reisman computation board	Same as IS 9.1.5 1.1.0 Rapid rate of learning	Same as IS 9.1.5
9.2.1 Emphasize patterns	Same as IS 9.1.1	Same as IS 9.1.1

Instructional Strategy	Related Generic Influence on Learning	Mathematical Topic Particularly Appropriate
9.2.2 Point out relevant relationships	1.9.0 Ability to attend to salient aspects of a situation 2.1.0 Disorders of visual perception 2.2.0 Disorders of auditory perception 3.3.0 Sensory limitation	5.1.0 Sequencing 5.2.0 One-to-one correspondence 5.3.0 Topological relationships 6.1.0 Duality 6.2.0 Equivalence of sets 6.3.0 Greater-less than 6.4.0 Sorting 6.5.0 Many-to-one and one-to-many matching 6.6.0 Seriation 6.7.0 Judging and estimating 8.1.0 Transformations 8.2.0 Set operations and relationships 8.3.0 Number operations and relationships 8.4.0 Mathematics as a language of relationships 9.2.0 Axioms 9.15.0 Economics
9.2.3 Use of modified conservation task	1.8.0 Ability to form relationships, concepts, and generalizations 1.9.0 Ability to attend to salient aspects of a situation 1.10.0 Use of problem-solving strategies 1.11.0 Ability to make decisions and judgments 1.12.0 Ability to draw inferences and conclusions	8.1.0 Transformations 9.2.0 Axioms 9.14.0 Applied mathematics—measurement
9.3.1 Use R-K counting boards: "Sequence" model for writing numerals up to 19	Same as IS 9.1.5	Same as IS 9.1.5

Instructional Strategy	Related Generic Influence on Learning		Mathematical Topic Particularly Appropriate	
9.3.2 Use R-K counting boards for writing numerals greater than 19		Same as IS 9.1.5		Same as IS 9.1.5
9.3.3 Use R-K counting boards: "Exchange" model for writing numerals greater than 9		Same as IS 9.1.5		Same as IS 9.1.5
9.3.4 Use Reisman computation boards	1.1.0	Rapid rate of learning Same as IS 9.1.5		Same as IS 9.1.5
9.4.1 Use R-K counting boards		Same as IS 9.1.5		Same as IS 9.1.5
9.5.1 Provide necessary structure	1.3.0	Ability to retain information	5.1.0	Sequencing
	4.1.0	Hyperactivity, distractibility, impulsivity	6.6.0	Seriation
			6.7.0	Judging and estimating
			9.5.0	Seriation extended
	4.2.0	Aggression		
	4.3.0	Withdrawal, immaturity, inadequacy		
	4.4.0	Deficiencies in moral development		
9.5.2 Use students' questioning skills	1.1.0	Rate and amount of learning	5.1.0	Sequencing
	1.5.0	Verbal skills	5.2.0	One-to-one correspondence
	1.8.0	Ability to form relationships, concepts, and generalizations	5.3.0	Topological relationships
			6.4.0	Sorting
	1.9.0	Ability to attend to salient aspects of a situation	6.6.0	Seriation
			6.7.0	Judging and estimating
			7.1.0	Concepts
	1.10.0	Use of problem-solving skills	8.4.0	Mathematics as a language of relationships
	1.12.0	Ability to make inferences and to hypothesize	9.5.0	Seriation extended
			9.6.0	Solving word problems
	4.4.0	Deficiencies in moral development	9.7.0	Cause–effect extended
			9.14.0	Applied mathematics—measurement

Instructional Strategy	Related Generic Influence on Learning	Mathematical Topic Particularly Appropriate
9.5.3 Provide both verbal-logical and visual-pictorial tasks	1.8.0 Ability to form relationships, concepts, and generalizations 1.9.0 Ability to attend to salient aspects of a situation	9.5.0 Seriation extended
9.6.1 Present problems with an unstated problem	1.10.0 Use of problem-solving strategies	9.6.0 Solving word problems
9.6.2 Present problems with incomplete information	Same as IS 9.6.1	Same as IS 9.6.1
9.6.3 Use problems with surplus information	Same as IS 9.6.1	Same as IS 9.6.1
9.7.1 Present problems with several solutions	Same as IS 9.6.1	Same as IS 9.6.1 9.7.0 Cause–effect extended
9.7.2 Present problems in a series involving several operations	Same as IS 9.6.1	Same as IS 9.6.1 9.7.0 Cause–effect extended
10.1.1 Use structured algorithms	1.13.0 Ability in general to abstract and cope with complexity 2.1.0 Disorders of visual perception 4.1.0 Hyperactivity, distractibility, impulsivity	9.4.0 Computation
10.1.2 Use hand calculator	1.1.0 Rate and amount 1.2.0 Speed of learning/content 1.3.0 Ability to retain information 1.4.0 Need for repetition 1.6.0 Ability to learn arbitrary associations and symbol systems 2.1.5 Difficulties in spatial relationships 3.1.0 Physical impairments 3.2.0 Low-vitality and fatigue conditions 3.3.0 Sensory limitation	9.4.0 Computation

Instructional Strategy	Related Generic Influence on Learning		Mathematical Topic Particularly Appropriate	
10.1.3 Restrict number of computations		Same as IS 10.1.2	9.1.0	Basic facts
			9.4.0	Computation
10.1.4 Develop a time schedule for completing a task	1.1.0	Rate and amount		All topics
	1.2.0	Speed of learning/content		
	1.10.0	Use of problem-solving strategies		
	1.11.0	Ability to make decisions and judgments		
	3.1.0	Physical impairments		
	3.2.0	Low-vitality and fatigue conditions		
10.1.5 Use tactile experiences	2.1.0	Disorders of visual perception	6.7.0	Judging and estimating
	3.3.0	Sensory limitation	8.1.0	Transformations
10.1.6 Use fixed representations of number	1.6.0	Ability to learn arbitrary associations and symbol systems	7.1.0	Concepts
	2.1.0	Disorders of visual perception		
	3.3.0	Sensory limitation		
11.1.1 Use cues	1.3.0	Ability to retain information	5.2.0	One-to-one correspondence
	1.9.0	Ability to attend to salient aspects of a situation	5.3.0	Topological relationships
	2.1.0	Disorders of visual perception	5.4.0	Arbitrary associations
	2.2.0	Disorders of auditory perception	6.3.0	Greater-less than
	3.3.0	Sensory limitation	6.6.0	Seriation
	4.1.0	Hyperactivity, distractibility, impulsivity	8.2.0	Set operations and relationships
	4.4.0	Deficiencies in moral development	8.3.0	Number operations and relationships
11.1.2 Use Fitzgerald Key	3.3.0	Sensory limitation	6.3.0	Greater-less than
			8.4.0	Mathematics as a language of relationships
			9.6.0	Solving word problems

Instructional Strategy	Related Generic Influence on Learning		Mathematical Topic Particularly Appropriate	
11.1.3 Use logographs	1.5.0	Verbal skills	8.4.0	Mathematics as a language of relationships
	1.6.0	Ability to learn arbitrary associations and symbol systems		
11.1.4 Use classroom management techniques	4.1.0	Hyperactivity, distractibility, impulsivity		All topics
	4.2.0	Aggression		
	4.3.0	Withdrawal, immaturity, inadequacy		
	4.4.0	Deficiencies in moral development		
11.1.5 Use multisensory approach	1.8.0	Ability to form relationships, concepts and generalizations		All topics
	1.13.0	Ability in general to abstract and cope with complexity		
	2.1.0	Disorders of visual perception		
	2.2.0	Disorders of auditory perception		
	3.3.0	Sensory limitation		
11.1.6 Use media		Same as IS 11.1.5		All topics
11.1.7 Use familiarity	1.3.0	Ability to retain information	5.2.0	One-to-one correspondence
	1.8.0	Ability to form relationships, concepts, and generalizations	6.2.0	Equivalence of sets
			6.4.0	Sorting
			9.6.0	Solving word problems
	3.3.0	Sensory limitation		
	4.3.0	Withdrawal, immaturity, inadequacy		
11.2.1 Use vision aids	3.3.0	Sensory limitation	5.4.0	Arbitrary associations
			9.3.0	Place value
			9.4.0	Computation
			9.6.0	Solving word problems

293

	Instructional Strategy	Related Generic Influence on Learning		Mathematical Topic Particularly Appropriate	
11.2.2	Use nonvisual cues	3.3.0	Sensory limitation	5.1.0 5.3.0 6.4.0	Sequencing Topological relationships Sorting
11.2.3	Use Braille numerals	3.3.0	Sensory limitation	9.1.0 9.4.0	Basic facts Computation
11.3.1	Minimize manual dexterity in performance requirements	3.1.0	Physical impairments	9.4.0 9.6.0	Computation Solving word problems
11.3.2	Adjust the pace of instruction	1.1.0 1.2.0 1.3.0 1.4.0 1.6.0 3.1.0 3.2.0	Rate and amount Speed of learning/content Ability to retain information Need for repetition Ability to learn arbitrary associations and symbol systems Physical impairments Low-vitality and fatigue conditions		All topics
11.3.3	Use team taking of tests		Appropriate for all students		All topics
12.1.1	Use behavior modification techniques	4.1.0 4.2.0 4.3.0 4.4.0	Hyperactivity, distractibility, impulsivity Aggression Withdrawal, immaturity, inadequacy Deficiencies in moral development		All topics
12.1.2	Provide a structured environment		Same as IS 12.1.1		All topics
12.1.3	Use self-instruction	1.9.0 2.0.0 4.1.0	Ability to attend to salient aspects of a situation Psychomotor abilities or motor effects of mental processes Distractibility		All topics

Instructional Strategy	Related Generic Influence on Learning		Mathematical Topic Particularly Appropriate	
12.1.4 Eliminate irrelevant or potentially distracting stimuli from instructional materials	4.1.0	Hyperactivity, distractibility, impulsivity	All topics	
12.1.5 Use modeling	4.1.0	Hyperactivity, distractibility, impulsivity	6.7.0	Judging and estimating
	4.2.0	Aggression	8.4.0	Mathematics as a language of relationships
	4.3.0	Withdrawal, immaturity, inadequacy	9.6.0	Solving word problems
	4.4.0	Deficiencies in moral development	9.15.0	Applied mathematics—economics
12.2.1 Select successful mathematics task to obtain high success rates	4.2.0	Aggression	All topics	
	4.3.0	Withdrawal, immaturity, inadequacy		
	4.4.0	Deficiencies in moral development		
12.3.1 Arrange appropriate environmental conditions	4.1.0	Hyperactivity, distractibility, impulsivity	All topics	
	4.2.0	Aggression		
	4.3.0	Withdrawal, immaturity, inadequacy		
	4.4.0	Deficiencies in moral development		
12.3.2 Teach recognition of important social cues		Same as IS 12.3.1	9.14.0	Applies mathematics—measurement
			9.15.0	Economics
12.3.3 Insure success in tasks		Appropriate for all students	All topics	
12.4.1 Sensitize students to moral issues	4.4.0	Deficiencies in moral development	9.13.0	Probability, statistics, graphs
			9.14.0	Applied mathematics—measurement
			9.15.0	Economics

H Diagnostic Guide By Mathematics Topic

Mathematics Topic	Related Special Educational Needs	Suggested Instructional Strategy (IS)
5.1.0 Sequencing	1.1.0 Rate of learning slower than age peers	5.1.1 Present small amounts of sequence; chunking
	1.3.0 Difficulty retaining information	5.1.2 Incorporate redundancy
	1.4.0 Needs repetition	5.1.3 Use of cues
	1.8.0 Difficulty forming relationships	5.1.4 Separate and under-line
	1.9.0 Inability to notice salient aspects of situations	5.1.5 Control number of dimensions that define a linear sequence
	1.13.0 Difficulty coping with complexity	5.1.6 Emphasize patterns
	2.1.0 Visual perception problems	5.1.7 Introduce additional relevant dimensions for generating an alternating pattern
	2.2.0 Auditory perception problems	5.1.8 Use rehearsal strategies
5.2.0 One-to-one corre-spondence	1.1.0 Rate of learning slower than age peers	5.2.1 Employ frequency to develop number word–object associations
	1.3.0 Difficulty retaining information	5.2.2 Employ familiarity to develop number word–object associations
	1.4.0 Need for repetition	
	1.5.0 Immature verbal skills	

Mathematics Topic	Related Special Educational Needs	Suggested Instructional Strategy (IS)
5.2.0 One-to-one corre-spondence contd.	1.6.0 Difficulty learning arbitrary associations 1.8.0 Difficulty learning basic relationships 1.9.0 Inability to attend to salient aspects of situations 2.0.0 Psychomotor diffi-culties 3.3.0 Sensory limitation 4.1.0 Distractibility	5.2.1 Employ frequency to develop number word–object asso-ciations 5.2.2 Employ familiarity to develop number word-object asso-ciations 5.2.3 Employ vividness or imagery to develop number word–object associations 5.2.4 Emphasize patterns that generate the sequences to be matched 11.1.7 Use familiarity 12.1.4 Eliminate irrelevant stimuli from instruc-tional materials
5.3.0 Topological relationships	1.8.0 Difficulty forming relationships 1.9.0 Difficulty attending to salient aspects of situations 2.0.0 Psychomotor diffi-culties 3.1.0 Physical impairment 3.3.0 Sensory limitation	5.3.1 Reinforce attention to a relevant dimension 5.3.2 Point out relevant relationship 5.3.3 Incorporate move-ment to develop order 11.1.5 Use multisensory approaches 11.2.2 Use nonvisual cues

Mathematics Topic	Related Special Educational Needs	Suggested Instructional Strategy (IS)
5.4.0 Arbitrary associations	1.1.0 Rate of learning slower than age peers 1.3.0 Retention difficulty 1.4.0 Needs repetition 1.5.0 Immature verbal skills 1.6.0 Difficulty learning symbol systems 1.7.0 Minimal vocabulary skills 1.9.0 Difficulty attending to salient aspects of a situation 2.0.0 Psychomotor difficulties 3.3.0 Sensory limitation	5.4.1 Emphasize differences in distinctive features of stimuli 5.4.2 Control irrelevant stimuli 5.4.3 Use the student's preferred dimension 5.4.4 Place stimuli to be discriminated in close proximity on initial tasks 5.4.5 Increase the number of relevant dimensions 5.4.6 Use cues that are different during initial learning 5.4.7 Include identical stimuli to serve as background and thus to aid attention to relevant cues 5.4.8 Reinforce alternative positions on tasks involving random response patterns 8.4.4 Use relational rebus symbols 11.1.1 Use cues 11.1.2 Use Fitzgerald Key 11.1.3 Use logographs 11.2.1 Use vision aids 11.2.3 Use Braille numerals
6.1.0 Duality	1.1.0 Rate of learning slower than age peers 1.8.0 Difficulty forming relationships 1.9.0 Difficulty attending to salient aspects of a situation 1.11.0 Difficulty making decisions and judgments 2.0.0 Psychomotor difficulties 3.3.0 Sensory limitation	6.1.1 Employ familiarity to develop notion of duality 6.1.2 Point out relevant relationships 6.1.3 Use redundancy and point out relevant dimension 6.3.1 Restrict complexity

Mathematics Topic		Related Special Educational Needs		Suggested Instructional Strategy (IS)	
6.2.0	Equivalence of sets	1.8.0	Difficulty forming relationships, simple generalizations	6.2.1	Control the number of dimensions embedded in the stimuli
		1.9.0	Difficulty attending to salient features	6.2.2	Use chunking to control distractions in the ground
		1.11.0	Difficulty making judgments	6.2.3	Prevent incorrect responses
		2.1.0	Disorders of visual perception		
6.3.0	Greater than–less than matching	1.8.0	Difficulty forming relationships and generalizations	6.3.1	Restrict complexity of incidental learning to the gross and obvious
		1.9.0	Difficulty attending to salient aspects	6.3.2	Encourage consistent responses to similar stimuli
		1.10.0	Lack of problem-solving strategies	6.3.3	Replace incidental learning tasks with structured intentional learning tasks
		1.11.0	Difficulty making judgments		
		1.13.0	Difficulty coping with complexity		
		2.1.0	Disorders of visual perception		
6.4.0	Sorting	1.1.0	Slower rate of learning	6.4.1	Reduce complexity of task
		1.8.0	Difficulty forming relationships	6.4.2	Use the familiar and useful
		1.9.0	Inability to notice salient aspects	6.4.3	Use the strategy of becoming more familiar with the familiar
		1.10.0	Poor problem-solving strategies		
		1.11.0	Difficulty making judgments		
		1.13.0	Difficulty coping with complexity		
		2.1.0	Disorders of visual perception		
		3.3.0	Sensory limitation		
		4.1.0	Distractibility		

	Mathematics Topic		Related Special Educational Needs		Suggested Instructional Strategy (IS)
6.5.0	Many-to-one and one-to-many matching	1.1.0	Slower rate of learning	6.5.1	Use consistent vocabulary
		1.2.0	Speed of learning affected by content of task	6.5.2	Restrict complexity of the task and incorporate preference into stimuli
		1.8.0	Difficulty forming relationships and generalizations		
		1.9.0	Difficulty attending to salient aspects of a situation		
		1.10.0	Poor problem solving		
		1.11.0	Difficulty making decisions and judgments		
		1.13.0	Difficulty coping with complexity		
		2.1.0	Disorders of visual perception		
		2.3.0	Difficulty with rules of general language and mathematics		
		3.3.0	Sensory limitation		
6.6.0	Seriation	1.1.0	Slower rate of learning	6.6.1	Use a model whose competence in task has been established
		1.2.0	Speed of learning effected by content	6.6.2	Use of verbalization by model
		1.3.0	Difficulty with retention	6.6.3	Control the number of dimensions embedded in the stimuli
		1.8.0	Difficulty forming relationships and generalizations	6.7.2	Provide sequencing experiences comprised of more abstract and/or subtle nature
		1.9.0	Difficulty attending to salient aspects		
		1.10.0	Poor problem-solving strategies		
		1.12.0	Difficulty making inferences		
		1.13.0	Difficulty abstracting and dealing with complexity		
		2.1.0	Disorders of visual perception		
		2.2.0	Disorders of auditory perception		
		2.3.0	Language disabilities		
		3.3.0	Sensory limitation		
		4.1.0	Distractibility		

Mathematics Topic	Related Special Educational Needs	Suggested Instructional Strategy (IS)
6.7.0 Judging and estimating	1.1.0 Slower rate of learning 1.8.0 Difficulty forming relationships and generalizations 1.9.0 Difficulty attending to salient aspects 1.11.0 Difficulty making judgments 2.0.0 Psychomotor difficulties 4.4.0 Deficiencies in moral development	6.7.1 Restrict complexity of the stimuli 6.7.2 Provide sequencing experiences comprised of more abstract and/or subtle nature 6.7.3 Encourage deferred judgment during problem-solving activity
7.1.0 Concepts	1.8.0 Difficulty forming concepts 1.9.0 Difficulty attending to salient aspects of a situation 1.10.0 Poor problem-solving skills 1.11.0 Difficulty making judgments 1.12.0 Difficulty drawing inferences 1.13.0 Difficulty abstracting and coping with complexity 2.0.0 Psychomotor difficulties 3.3.0 Sensory limitation	7.1.1 Use of familiar categories to insure meaning for the student 7.1.2 Provide bimodal input 7.1.3 Group input by category 7.1.4 Increase task complexity 7.1.5 Employ creative problem-solving strategy to facilitate concept formation 8.2.2 Use natural categories for classification tasks 8.2.3 Distinguish between intensional and extensional aspects of concepts 8.2.4 Distinguish between classes and collections 8.2.5 Use good exemplars of a class 8.2.6 Restrict the number of exemplars in classification tasks
8.1.0 Transformations	Same as for 6.6.0 Seriation	8.1.1 Use modified conservation tasks 8.1.3 Apply modified conservation tasks to rigid transformations 8.1.4 Employ creative problem solving

301

Mathematics Topic	Related Special Educational Needs	Suggested Instructional Strategy (IS)
8.2.0 Set operations and relationships	Same as for 6.6.0 Seriation	8.2.1 Control for component skills 8.2.2 Use natural categories for classification tasks rather than arbitrary categories 8.2.3 Distinguish between intensional and extensional aspects of concepts 8.2.4 Distinguish between classes and collections 8.2.5 Use good exemplars of a class 8.2.6 Restrict the number of exemplars in classification tasks
8.3.0 Number operations and relationships	Same as for 6.6.0 Seriation	8.3.1 Provide immediate knowledge of results 8.3.2 Use distributed repetition 8.3.3 Use verbal rehearsal 8.3.4 Facilitate attention to the relevant dimension 8.3.5 Plan for transfer in learning 8.3.6 Employ Torrance's Hierarchy of Creative Skills

Mathematics Topic	Related Special Educational Needs		Suggested Instructional Strategy (IS)	
8.4.0 Mathematics as a language of relationships	1.5.0	Immature verbal skills	8.4.1	Use shorter, simpler sentences when giving directions
	1.6.0	Inability to learn symbol systems	8.4.2	Use concrete examples of spatial and quantitative relationships
	1.7.0	Size of vocabulary meager in relationship to that of age peers		
	1.8.0	Difficulty forming relationships, concepts, and generalizations	8.4.3	Use prompting
			8.4.4	Use relational rebus symbols
	1.9.0	Difficulty attending to salient aspects	8.4.5	Incorporate modeling plus verbal instructions
	1.10.0	Poor problem-solving skills		
	1.12.0	Difficulty hypothesizing	8.4.6	Highlight inflectional endings
	1.13.0	Difficulty abstracting	8.4.7	Introduce novelty into learning experience
	2.1.0	Disorders of visual perception		
	2.2.0	Disorders of auditory perception	8.4.8	Show how different verbal directions may change spatial relationships
	2.3.0	Difficulty with rules of language		
	3.3.0	Sensory limitation		
9.1.0 Basic facts	1.1.0	Slower rate of learning	9.1.1	Emphasize patterns
	1.3.0	Retention problems	9.1.2	Use rehearsal strategies
	1.4.0	Needs repetition		
	1.8.0	Difficulty learning concepts and generalizations	9.1.3	Emphasize differences in distinctive features of stimuli
	1.9.0	Difficulty attending to salient aspects	9.1.4	Restrict size of sum
	1.13.0	Difficulty abstracting		
	2.2.0	Disorders of auditory perception		
	3.3.0	Sensory limitation		

	Mathematics Topic	Related Special Educational Needs		Suggested Instructional Strategy (IS)	
9.1.0	Basic facts contd.			9.1.5	Use R-K counting board
				9.1.6	Make two-for-one exchanges on R-K counting board
				9.1.7	Use Reisman computation board
9.2.0	Axioms		Same as for 9.1.0 Basic Facts	9.2.1	Emphasize patterns
				9.2.2	Point out relevant relationships
				9.2.3	Use of modified conservation task
9.3.0	Place value	1.1.0	Slower rate of learning	9.3.1	Use R-K counting boards: sequence model for numerals up to 19
		1.2.0	Speed of learning affected by learning task topic		
		1.3.0	Retention problems	9.3.2	Use R-K counting boards for writing numerals greater than 19
		1.4.0	Needs repetition		
		1.8.0	Difficulty forming relationships, concepts, and generalizations		
		1.9.0	Inattentive to salient aspects	9.3.3	Use R-K counting boards: Exchange model for writing numerals greater than 9
		1.10.0	Poor problem-solving skills		
		1.11.0	Difficulty making judgments	9.3.4	Use Reisman computation boards
		1.12.0	Difficulty inferring		
		1.13.0	Difficulty dealing with complexity		
		2.0.0	Psychomotor difficulties		
		3.3.0	Sensory limitation		

Mathematics Topic	Related Special Educational Needs	Suggested Instructional Strategy (IS)
8.4.0 Computation	Same as for 8.3.0 Place value 11.3.0 Physical impairment	9.4.1 Use R-K counting boards 10.1.1 Use structured algorithms 10.1.2 Use hand calculator 10.1.3 Restrict numbers of computations 11.2.3 Use Braille numerals 11.3.1 Minimize manual dexterity in computations
9.5.0 Seriation extended	Same as for 8.3.0 Place value	5.1.6 Emphasize patterns that generate the sequence 5.1.7 Introduce additional relevant dimensions for generating an alternating pattern 5.1.9 Provide sequencing activities comprised of more complex or subtle nature 5.1.10 Incorporate incompleteness to activate creative potential 6.7.2 Provide sequencing experiences comprised of more abstract and/or subtle nature 8.3.4 Facilitate attention to the relevant dimension 9.5.1 Provide necessary structure 9.5.2 Use students' questioning skills 9.5.3 Provide both verbal-logical and visual-pictorial tasks

Mathematics Topic	Related Special Educational Needs		Suggested Instructional Strategy (IS)	
9.6.0 Solving word problems	1.1.0	Slower rate of learning	5.1.10	Incorporate incompleteness to activate creative potential
	1.2.0	Speed of learning affected by topic	5.2.3	Employ vividness or imagery to represent the number word-object associations represented in word problems
	1.5.0	Immature verbal skills		
	1.6.0	Difficulty learning symbol systems		
	1.7.0	Minimal vocabulary		
	1.8.0	Difficulty forming relationships, concepts and generalizations	5.3.2	Point out relevant relationships
	1.9.0	Inattention to salient aspects	6.5.1	Use consistent vocabulary
	1.10.0	Poor problem-solving strategies	6.7.1	Restrict complexity
	1.12.0	Difficulty drawing conclusions	7.1.1	Use familiar categories
			7.1.4	Increase task complexity
	1.13.0	Difficulty abstracting and coping with complexity	8.3.1	Provide immediate knowledge of results
	2.1.0	Disorders of visual perception	8.3.4	Facilitate attention to the relevant dimension
	2.2.0	Disorders of auditory perception	8.4.2	Use concrete examples of spatial and quantitative relationships
	2.3.0	Difficulty with rules of general language and mathematics		
	3.3.0	Sensory limitation	8.4.5	Incorporate modeling
	4.2.0	Aggressiveness	8.4.6	Highlight inflectional endings
	4.3.0	Withdrawal, immaturity, inadequacy	9.6.1	Present problems with an unstated problem
			9.6.2	Present problems with incomplete information
			9.6.3	Use problems with surplus information
			9.6.4	Present problems with several solutions
			9.6.5	Present problems in a series involving several operations
			10.1.2	Use hand calculator
			11.1.2	Use Fitzgerald Key
			11.1.3	Use logographs
			11.1.7	Use familiarity

Mathematics Topic	Related Special Educational Needs	Suggested Instructional Strategy (IS)
9.6.0 Solving word problems contd.		11.2.1 Use vision aids
		11.3.1 Minimize manual dexterity in performance requirements
		11.3.3 Use team problem-solving
		12.1.5 Use modeling
		12.2.1 Select mathematics tasks to insure high success rates
		12.3.3 Insure success in tasks
9.7.0 Cause–effect extended	Same as for 9.6.0 Solving word problems	5.1.6 Emphasize patterns that generate a sequence
		5.1.7 Introduce additional relevant dimensions for generating an alternating pattern
		5.1.9 Provide sequencing activities comprised of more complex or subtle nature
		5.3.2 Point out relevant relationship
		8.1.1 Use modified conservation tasks
		9.5.2 Use students' questioning skills
9.8.0 Topological equivalence	Same as for 5.3.0 Topological relationships	5.1.10 Incorporate incompleteness to activate creative potential
		5.3.1 Reinforce attention to a relevant dimension
		5.4.1 Emphasize differences in distinctive features of stimuli
		8.1.1 Use modified conservation tasks
		8.1.3 Apply modified conservation tasks to rigid transformations
		8.4.2 Use concrete examples of spatial and quantitative relationships

	Mathematics Topic	Related Special Educational Needs	Suggested Instructional Strategy (IS)	
9.9.0	Projective geometry	Same as for 9.6.0 Solving word problems	8.1.4	Employ creative problem solving
			8.4.7	Introduce novelty into learning experience
9.10.0	Nonmetric geometry	Same as for 9.6.0 Solving word problems	5.3.1	Reinforce attention to a relevant dimension
			5.4.1	Emphasize differences in distinctive features of stimuli
9.11.0	Metric geometry	Same as for 9.6.0 Solving word problems	5.3.1	Reinforce attention to a relevant dimension
			6.3.1	Restrict complexity of incidental learning to the gross and obvious
			8.4.2	Use concrete examples of spatial and quantitative relationships
			10.1.2	Use hand calculator
			11.3.3	Use team taking of tests
			12.1.4	Eliminate irrelevant or potentially distracting stimuli from instructional materials
9.12.0	Network theory	Same as for 5.3.0 Topological relationships	5.3.1	Reinforce attention to a relevant dimension
			8.1.4	Employ creative problem solving
			8.2.5	Use good exemplars of a class
9.13.0	Probability, statistics, graphs	Same as for 9.6.0 Solving word problems	5.1.10	Incorporate incompleteness to activate creative potential
			7.1.4	Increase task complexity
			12.1.5	Use modeling

	Mathematics Topic	Related Special Educational Needs	Suggested Instructional Strategy (IS)	
9.14.0	Measurement— applied mathematics	Same as for 9.6.0 Solving word problems	5.3.2	Point out relevant relationships
			6.6.1	Use a model whose competence in the task has been established
			12.3.2	Teach recognition of important social cues
			12.3.3	Insure success in tasks
9.15.0	Economics	Same as for 9.6.0 Solving word problems	5.3.2	Point out relevant relationships
			6.7.3	Encourage deferred judgment during problem solving activity
			9.2.2	Point out relevant relationships
			12.1.5	Use modeling
			12.3.2	Teach recognition of important social cues
			12.4.1	Sensitize students to moral issues

Author Index

Abercrombie, M. L. J., 21, 29
Abeson, A., 34, 35
Adcock, C., 15
Aiello, B., 33
Aiken, L. R. Jr., 28
Albarn, K., 73, 91
Altman, R., 91, 92
Alves, S., 187, 195
Annesly, F. R., 43
Arieti, S., 42, 43, 91, 92, 99
Ashcroft, M. H., 139
Association for Childhood Education
    International, 43, 44

Bandura, R. F., 41, 43, 200
Bannatyne, A., 10, 15
Barraga, N., 2
Beedle, R. K., 67, 71
Bellugi, U., 6, 15
Belmont, L., 25, 29
Bemporad, J., 42
Bender, L., 42
Bentzen, F., 197, 201
Bettelheim, B., 42
Biehler, R. F., 42, 43
Bigge, J., 22, 24, 28, 29, 30, 33, 34
Biggs, J. B., 40, 43
Bijou, S. W., 14
Billings, H. K., 33
Birch, H. G., 25, 29, 30
Birch, J. W., 34, 99, 195
Bish, C. E., 15
Bitter, G. B., 194
Blake, K. A., 52, 55, 61, 64, 65, 71, 91, 99
Blatt, B., 43

Bliss, C. K., 8, 15
Boatner, M. T., 194
Bolick, N., 34, 35
Bornstein, H., 192
Bracht, G. H., 15
Braggio, J. T., 159
Braggio, S. M., 159
Bricker, D. D., 70
Bricker, W. A., 70
Broen, P. A., 7, 10, 16
Brophy, J. E., 200
Brothers, R. J., 194
Brown, A. L., 68, 71
Brown, R., 6, 15
Buchanan, R., 33
Burton, T., 99

Caravella, G. R., 194
Carnine, D., 70
Carrow, M. A., 29
Casteneda, A., 43
Chomsky, C., 29
Christopher, D. A., 194
Clark, A., 187, 195
Clark, C. R., 7, 8, 9, 15
Clements, J. E., 200
Cohen, R., 29
Compton, C., 22, 24, 28, 29, 30, 34
Compton, M. F., 42
Connor, F. P., 31, 32, 34, 35
Crandall, V. C., 40, 44
Cronbach, L. J., 52, 71
Cruickshank, W. M., 29, 34, 43, 197, 201
Csapo, M., 198, 201
Cullinan, D. A., 198, 201

311

Dahmus, M. E., 29
Davidson, D. S., 40, 43, 44
Davis, H., 194
Dawes, C., 84, 91
Dennison, A. L., 195
Dunn, M., 4, 15

Eaton, W. O., 43
Ebmier, H., 200
Edwards, L. L., 42
Elkind, D., 91
Ellis, N. R., 56, 71, 99, 100
Engleman, S., 15, 70
Erikson, E. H., 44
Euphemia, R., 34
Evans, R. A., 67, 71

Farr, G., 159
Fernald, G. M., 70
Fisher, M. A., 63, 67, 68, 71
Fitzgerald, E., 185, 186, 195
Flavell, J. H., 71, 91
Fliegler, L. A., 15
Fodor, J. D., 15
Forman, G. E., 43
Foster, R., 38, 44
Frish, S. A., 15
Furth, H. G., 194

Gallagher, P. A., 42
Galvin, J. P., 43
Gates, J. E., 194
Gelb, I. J., 7, 15
Goettler, D. R., 71
Good, T. L., 200
Goodwin, S. E., 197, 198, 201
Goslin, D., 44
Gowan, J. C., 15
Gruber, H. E., 15, 17, 99
Guilford, J. P., 5, 15, 28, 29

Hadamard, J., 15
Hall, S. M., 92
Hallahan, D., 29, 43, 197, 201
Hamilton, L. B., 192, 195
Hanfmann, E., 96, 99, 100
Hanninen, K. A., 194

Haring, N. G., 43, 197, 201
Harris, L. J., 15, 16
Harris-Vanderheiden, D., 8, 16
Hartup, W. C., 44
Haskell, S. H., 31, 32, 34
Hass, J., 34, 35
Heady, J., 200
Hebb, D. O., 20, 29
Helfgott, S., 187, 195
Hetherington, E. M., 43
Hewett, F. M., 197, 201
Hill, K. P., 43
Hoch, P. H., 43
Hood, H. B., 4, 15
Horowitz, R., 29
House, B. J., 65, 71
Howard, R. L., 192
Hunt, D., 187, 195

Inhelder, B., 4, 16

Johnson, G. O., 34
Johnson, N. S., 139
Johnston, R. B., 29, 34
Junkala, J. B., 42

Kaliski, L., 29
Kamii, C. K., 15
Kappan, D. L., 195
Kasanin, J., 96, 99, 100
Kauffman, J. M., 40, 44, 196, 197, 198, 199, 201
Kaufman, A. S., 5, 16
Kellas, G., 139
Kilpatrick, J., 44, 71, 86, 91, 99
Kohlberg, L., 44, 200, 201
Koupernik, C., 21, 29
Kruschner, D. S., 43
Krutetskii, V. A., 16, 17, 38, 44, 70, 71, 86, 91, 99, 166, 167, 168, 169

La Fleur, N. K., 198, 201
Laurendeau, M., 91
Lehtinen, L. E., 20, 29
Leland, H., 38, 44
Lighthall, F. F., 40, 44
Linville, W. J., 29

Lipsett, L. P., 71
Litrownik, A. J., 91, 92
Lovell, K., 99
Lowenfeld, B., 195
Luria, A. R., 16, 17, 18, 19, 197, 201

Mackeith, R., 21, 29
Magrab, P. R., 29, 34
Mahoney, M. J., 197, 198, 201
Markman, E. M., 201
McCandless, B. R., 43
McClelland, D., 15
McDonnell, F., 195
McNeil, D., 29
Mears, E. G., 194
Meeker, M. N., 5, 16
Menyuk, P., 29
Mercer, C. D., 54, 56, 57, 66, 67, 68, 71, 72,
    74, 87, 91, 99
Merrill, M. A., 5, 16
Meyer, W. J., 43
Meyers, R. E., 99
Miller, R. R., 192
Mullins, J. B., 30, 33, 34
Myklebust, H. R., 37, 44, 195

Napier, G., 195
Nhira, K., 38, 44
Niewoehner, M., 200
Nober, L., 187, 195
Norman, D. A., 56, 71
Northcott, W. H., 195

O'Donnell, P. A., 33, 34
Osborn, D. K., 43
Osborn, J. B., 43
Ottina, J. R., 34

Palermo, D. S., 43
Papy, F., 146
Pasanella, A. L., 43
Paul, J. L., 42
Petrie, S., 187, 195
Phillips, C. J., 31, 34
Phillips, E. L., 197, 201
Pinard, A., 91
Platt, J., 200

Ratzeburg, F., 197, 201
Reading, F. R., 43
Reisman, F. K., 38, 44, 62, 71, 84, 91, 146,
    159, 178, 179, 183
Resnick, L. B., 29
Ress, H. W., 71
Reusch, J., 43
Reynolds, C. R., 43
Reynolds, M. C., 30, 34, 99, 195
Richmond, B. D., 43
Rogers, L., 15
Rosenthal, D. D., 29
Ross, D., 91, 92
Routh, D. K., 68, 71
Ruebush, B. K., 40, 44

Sarason, I. G., 43
Sarason, S. B., 39, 40, 43, 44
Sattler, J. M., 31, 34, 71, 99
Sauliner, K. L., 192
Sax, G., 34
Scholl, G. T., 195
Schreibman, L., 201
Schrotberger, W. L., 195
Schultz, J. B., 172
Schumaker, J. B., 15
Semel, E., 28, 29
Shantz, C., 43
Shellhaas, M., 38, 44
Siegel, G. M., 7, 10, 16
Silverman, S. R., 194
Simpson, L., 159
Singer, R. V., 77, 91
Smith, J. M., 73, 91
Smith, V. H., 29
Smothergill, N. L., 44
Snell, M. E., 54, 56, 57, 66, 67, 68, 71, 72,
    74, 87, 91, 99
Snow, R. E., 52, 71
Spitz, H. H., 69, 71, 95, 99, 100
Stevens, G. D., 34, 35
Strauss, A. A., 20, 29
Strichert, S. S., 87, 92
Striffler, N., 27, 29, 32, 34
Szasz, T. S., 43

Talkington, L. W., 91, 92
Tannhauser, M. A., 197, 201
Teller, J., 44, 71, 86, 99

Terman, L. M., 5, 16
Tobias, S., 40, 44
Torrance, E. P., 59, 71, 82, 83, 90, 92, 96, 98, 99, 100
Turnbull, A. P., 172
Tuttle, D. W., 195

Ullman, D. G., 68, 71

Vanderheiden, G. C., 8, 16
Vanderlinde, L. F., 29
Volkmor, C. B., 43
Voneche, J. J., 15, 17, 99
Vurpillot, E., 93, 100
Vygotskii, L. S., 16, 17, 96, 100

Waite, R. R., 40, 44
Ward, P., 195

Watson, D., 195
Watson, T. J., 195
Waugh, N. C., 56, 71
Webb, T. E., 43
Webreck, C. A., 71
Wechsler, D., 5, 16
Weikart, D. P., 15
Weiner, S., 187, 195
Whelan, R. J., 197, 201
White, R. R., 31, 34
Wiig, E., 28, 29
Williams, J. F., 21, 29
Wilson, J. W., 4, 15
Wirszup, I., 44, 71, 86, 91, 99
Wood, M. M., 38, 41, 44, 201
Woodcock, R. W., 7, 8, 9, 15
Woodward, M., 4, 16

Zeaman, D., 63, 65, 67, 68, 71
Zubin, J., 43

Subject Index

AAMD Adaptive Behavior Scale, 38–39
Abstraction, 94
Addition, 231
  definition of, 117
  examples of, 144, 146
Aggressive children
  the teaching of, 197–198
Aggressiveness, 41
Anxiety, 39–40
Arbitrary associations, 64–69
Associative law for multiplication, 152
Associative property, 231
Asymmetry, 84–85
Auditory discrimination, 25
Auditory perception
  definition of, 24
  difficulty in sound blending, 25
  disorders of, 24
  figure-ground distractibility, 25
  poor auditory analysis, 25
  poor auditory discrimination, 25
  problems in sequential memory, 25
Auditory sequential memory, 25
Auditory synthesis, 24
Axioms, 148, 231, 304

Basic facts, 303–304
Basic-level classification, 113
Bliss symbols, 8–9
Braille code, 189–190

Cardinality, 105
Case studies, 258–269
Classes versus collections, 115

Classifying versus sorting, 81–82
Communication blocks, 207–209
Commutative law for multiplication, 151
Commutative property, 231
Composite number, 233
Computation, 159
  test of, 160
Concepts
  classes versus collections, 115
  definitions of, 94
  extensional versus intensional aspect of, 114
  intensional-extensional classification, 114
Conservation, 102–105
  as related to equilibration, 104–105
  as related to reversibility, 104
  in modified conservation tasks, 108
  use of, 105
Convergent-divergent thinking cycle, 108
Counting objects
  versus counting on number line, 254
Cue distinctiveness, 66–67
Cumulative rehearsal, 57
Curriculum, 45

Decimals, 234
Developmental Mathematics Curriculum
  Screening Checklist, 218, 220, 243–253, 260
Developmental Therapy Model, 41
Diagnostic guide
  by mathematics topic, 296–309
Diagnostic teaching, 226
Dichotomy, 73
Direct question, 206

Discrimination learning, 64–65
Distractibility, 40
Distributive property, 231
Division
   examples of, 145, 148
Duality, 73–74, 298

Economics, 309
Education of All Handicapped Children Act of
   1975, 30
The Education for All Handicapped Children
   Act (Public Law 94-142), 203, 212–214,
   226
   parents' rights under, 227
Emotional factors influencing learning
   mathematics, 37–42
   aggressiveness, 41
   anxiety, 39–40
   distractibility, 40
   impulsivity, 40
   moral development, 41–42
   withdrawal, 41
Equilibration, 104
Equivalence, 75–77
Equivalence of sets, 299
Equivalence relation, 233
Extension, definition by, 110

Fact-finding versus fact-giving interviews, 205
Figure-ground distractibility
   auditory, 25
   visual, 22–23
Fitzgerald Key, 185–186
Function, 232

General Anxiety Scale for Children, 40
Generalization
   definition of, 138
Geometry, 235, 308
Greater than – less than matching, 299
Greatest Common Factor (GCF), 234

Hearing-impaired, 184–186
   the teaching of, 186
Higher-level generalizations
   with advanced seriation, 165
   as axioms, 148–151

as basic facts, 138
   in computation, 159
   in place value, 152–159
   in solving word problems, 167
Higher-level relationships, 101
   conservation, 102–105
   mathematics as language, 123
   operations and relationships, 109–122
   transformations, 101–102
Hyperactive
   self-instruction by the, 197
   the teaching of the, 196–197
Hyperactivity (or hyperkinesis), 40

IEPM Implementation Plan, 221
IEPM Total Service Plan, 221
Implementation (of mathematics learning),
   203
   by developing IEMPs, 214
   by interviews, 205
   through parent aid, 226
   by use of PL 94-142, 212–214
Impulsivity, 40
Indirect question, 206
Individualized Education Programs in
   Mathematics (IEPMs), 203, 212
   annual goals of, 215
   examples of, 221–223, 259–269
   format of, 221
   functions of, 215
   in parent-teacher cooperation, 226
   purpose of, 214
Instructional Strategies (ISs)
   arbitrary associations, 64–69
   concepts, 93–98
   cross-reference outline of, 270–296
   higher-level generalizations, 138–172
   lower-level generalizations, 73–90
   one-to-one correspondence, 60–62
   pattern emphasis, 55, 139
   for psychomotor disabilities, 175–182
   rehearsal strategies, 56–58, 139
   for sensory and physical needs, 184–192
   sequencing, 52–53
   for social and emotional needs, 196–199
      (Also see Chapt. 4)
   topological relationships, 62–64
Instructional Strategies (IS) for Specific
   Learning Disabilities in Mathematics
   (SLDM) Checklist, 240–242

Intension, definition by, 110
Intensional-extensional classification, 114
Interviewing techniques, 205
   questioning skills, 206
Invariance (or invariant properties), 101
   as related to conservation, 102

Judging and estimating, 89–90

Leading question, 207
Least Common Multiple (LCM), 234
Lower-level generalizations, 73
   duality, 73–74
   equivalence, 75–77
   greater than–less than matching, 77–79
   judging and estimating, 89
   many-to-one and one-to-many matching, 83
   many-to-one multiplication, 153
   seriation, 84–88
   sorting, 80–83
Low-vitality and fatigue conditions
   in learning mathematics, 32

Mainstreaming, 30, 212
Manipulative interviews, 205–206
Many-to-one correspondence
   example of, 101, 146
Many-to-one and one-to-many matching, 300
Matching
   greater than–less than, 77–79
   many-to-one and one-to-many, 83–84
   equivalence, 84
   representation, 84
Matching graph, 84
Mathematics
   as language of relationships, 123
   as quantitative, spatial, and temporal
     relationships, 124, 127–128
Mathematics, learning
   and ability to form relationships, 10–11
   and ability to make judgments, conclusions,
     and abstractions, 13–14
   rate of, 3–5
   repetition of, 5–6
   retention of, 5
   and size of vocabulary, 10
   and symbol systems, 7–9

and use of problem solving, 12–13
   and verbal skills, 6–7
Measurement, 309
Memory, definition of, 5
Method of instruction, 45, 47
Methodology, 47
Modified conservation tasks, 108
   at the concrete-hypothetical level, 109
   at the concrete level, 109
   at the hypothetical-hypothetical level, 109
Moral development
   conventional, 41
   deficiencies in, 41
   postconventional, 41–42
   preconventional, 41
Morphological rules, 26
Multiplication, 231, 256, 257
   associative law, 152
   commutative law, 151
   definition of, 117
   examples of, 144, 147
   multiplicative identity, 151
   network method of, 178–179
   place value, 153
Multiplicative identity, 151

Numerical versus nonnumerical judgments, 110

One-to-many correspondence
   example of, 101
One-to-one correspondence, 60–62, 296–297
   word number-object associations, 61
Order, 121
Ordered pair, 232
Output, 21

Paired-associates learning, 61
Pattern, 122
Perception, 21, 32
Perceptual invariance versus conceptual
   invariance, 93
Phonological rules, 26
Physical impairments
   as influences in learning math, 31–32
Physically impaired, 189–190
   the teaching of the, 191
   use of team testing of, 191

Piagetian conservation tasks, 102
  as modified, 105–107, 108
Place value, 84, 152–158, 219, 255, 257, 304
Postconventional morality, 41–42
Preconventional morality, 41
Primary anxiety, 36
Prime number, 233
Probability, 236, 308
Psychomotor disabilities, 175
  as related to affect, 177
  as related to motor performance, 176
  as related to orientation, 176
  as related to perception, 176
  as related to problem solving, 176
  as related to reasoning, 175
Psychomotor learning
  effects on learning mathematics, 20–30
  three stages of, 21
Public Law 94-142 (The Education for All Handicapped Children Act), 203–214, 226, 227

Questions, types of, 206

Randomness, 122
Rate of mathematics learning, 3
  accelerated, 4
  retarded, 4
Real number system, 148, 233
Rehearsal strategies, 56–58
  cumulative, 57
  scanning and pointing, 57
  verbal elaboration, 57
Reisman computation board
  use of, 146
Reisman-Kauffman Interview Checklist, 237–239
Relationships, 303
  reflexive, 232
  symmetric, 232
  transitive, 232
Representation, 84
Reversibility, 104

Scanning and pointing, 57
Secondary anxiety, 36
Semantical rules, 27

Sensory awareness
  definition of, 32
Sensory integration, 25
Sensory limitation
  as related to perception of math, 32–33
Sensory and physical needs
  impaired hearing, 184–86
  impaired vision, 188
  low-vitality and fatigue conditions, 192
  physical impairment, 189
Sequencing, 52–53, 296
  auditory, 54
  examples of, 53–60
  visual, 54–55
Sequential memory
  auditory, 25
  visual, 23
Seriation, 84–88, 300
  advanced, 165
  extended, 305
Series, 52
Set, definition of, 110
Social factors influencing learning
  attention, 37–38
  completion of assignments, 38
  cooperation, 37
  mathematics, 37–38
  organization, 38
  responsibility, 38
  social acceptance, 38
  tactfulness, 39
Sorting, 80–83
  versus classifying, 81–82
Spatial relationships
  auditory, 25
  visual, 23–24
Specific Learning Disabilities in Mathematics (SLDM) Checklist, 177–178, 240–242
Subtraction, examples of, 144, 147
Succession, 52
Superordinate classification, 114
Symbol systems, 7–9
Symptomatic anxiety, 36–37
Synonym, 101
Syntactical rules, 27–28

Team testing, 191
Test Anxiety Scale for Children, 39
Time, how to teach, 256
Topological equivalence, 307–308

318

Topological relationships, 62–64
  enclosure, 63–64
  order, 64
  proximity, 62–63
  separateness, 63–64
Transfer (in learning), 120
Transformation, 101, 301
  as related to conservation, 102, 105
Transitivity, 85
Treatment interviews, 206

Verbal elaboration, 57
Vision-impaired, 188
  the teaching of, 189
Visual discrimination, 22
  figure-ground distractibility, 22–23

Visual discrimination (Cont.)
  form-constancy problems, 23
Visual perception
  disorders of, 22
  figure-ground distractibility, 22–23
  form-constancy problems in, 22–23
  low visual-sequential memory in, 22–23
  spatial relationships in, 23–24
Visual-spatial organization, 20
Visuomotor disorders, 20–21

Word problems, 306

Zero, discrepancy of, 255

DATE DUE

CO 38-297